LABORATORY METHODS

ENCYCLOPAEDIA OF CELL BIOLOGY - 4

LABORATORY METHODS

By

Dr. M. Prakash

Dept. of Zoology
M.M.H. Post Graduate College
Ghaziabad (U.P.)
(India)

DISCOVERY PUBLISHING HOUSE PVT. LTD.
NEW DELHI-110 002

Published by:
Namit Wasan

DISCOVERY PUBLISHING HOUSE PVT. LTD.
4383/4B, Ansari Road, Darya Ganj
New Delhi-110 002 (India)
Phone : +91-11-23279245; 23253475; 43596065
E-mail : discoverybooksindia@gmail.com
discoverypublishinghouse@gmail.com
namitwasan9@gmail.com
web : www.discoverypublishinggroup.com

***First Edition:* 2010**

***Reprinted:* 2020**

ISBN: 978-81-8356-454-0 (Set)
ISBN: 978-81-8356-562-2

Laboratory Methods

Printed at:
Infinity Imaging Systems
Delhi

Preface

The present title "Encyclopaedia of Cell Biology" has been written for undergraduate and postgraduate students of all Indian universities. It is a text on structural and functional unit of the organism that has been thoroughly checked and ornamented with easy and clear illustrations. Special emphasis has been concentrated on the fundamentals of the subject. All related students will find this well established text eminently suitable for introductory courses. Professional cell biologists, researchers, and students may find it useful for updating, and reference purposes. One of our major concerns of this title is to provide students with a basic understanding of what a cell is and why the cell is the fundamental unit of life. Our approach throughout has been to focus on the major question involved and the experimental approaches utilized in addressing these questions. Each important point is illustrated with an example to help clarify the idea; the numerous line drawings have all been chosen or created to help convey particular points. The result is a text that is easy to read, without resorting to verbal gimmicks or talking down to readers.

To make the work more comprehensive and informative, the author has consulted many authoritative books, research journals, abstracts, monographs etc., so there can be no claim to originality except in the manner of treatment.

The author expresses his thanks to his friends and colleagues whose continue inspirations have initiated him to bring out this book.

The author expresses his gratitude to Mr. Wasan and staff of M/s Discovery Publishing House Pvt. Ltd. for their whole hearted cooperation in the publication of this book.

Author

Preface

Contents

1

INTRODUCTION

Histochemical techniques enable the identification and localisation of specific substances within tissues. The methods depend on chemical reactions between the substance to be identified and localised in a tissue section, and one or more *reagents* in which the tissue section is incubated. The histochemist tries to arrange matters so that the end product of the chemical reaction is both coloured and insoluble, and therefore easily visible on microscopy. *Histochemical* methods such as the PAS reaction and the Perls' Prussian Blue reaction are well known to histopathologists, but until recently the scope of *histochemistry* has been limited by one big problem.

To cut thin sections of tissue it is necessary for it to be made rigid and to be embedded in a firm supporting medium; the embedding medium used in routine histopathology is *paraffin wax*. The process of embedding the tissue is a violent one, involving the immersion of the tissue in hot molten wax; it is essential, to minimise the distortion and disruption of the tissue during processing, that the tissue is adequately fixed beforehand. Furthermore the tissue must be completely *dehydrated* with alcohol and finally infiltrated with an organic solvent (usually toluene) with which the embedding wax is totally miscible.

The major limiting factor in histochemistry is that all these processes, so essential for the successful cutting of thin sections, inevitably chemically modify or completely destroy or remove the very substances we are attempting to *demonstrate*. Lipids in the tissue are dissolved out by the alcohols and xylene used in dehydration and clearing of the tissue. Many tissue enzymes are destroyed by the fixative and by the heat necessary in the embedding of the tissue in wax, and the accurate *intracytoplasmic localisation* of soluble substances such as glycogen is impossible in tissue exposed to aqueous fixatives.

Tissue to be investigated histochemically must therefore be processed in an entirely different manner to tissue for normal *histological stains*, although certain very stable substances can be demonstrated by histochemical means in paraffin sections. These are discussed in other chapter of this book. For the majority of histochemical reactions, however, an entirely new method of tissue processing and section cutting has had to be devised, and most of the improvements in histochemistry have only emerged in recent years after the development of satisfactory methods of tissue preparation.

The most important problem to overcome was that of producing high quality thin sections of tissue without the use of a supporting

Table 1.1 : Different methods of tissue processing and their application to histochemical procedures.

	Processing method	*Histochemical method*
(1)	Routine paraffin processing	Carbohydrates Proteins Pigments
(2)	Special paraffin processing	Proteins Hydrolytic enzymes (if no other method available)
(3)	Freeze drying	Carbohydrates Proteins
(4)	Freeze drying formol-vapour fixation	Carbohydrates Proteins Fluorescence techniques
(5)	Freeze drying with double embedding	Hydrolytic enzymes Proteins Carbohydrates
(6)	Frozen sections formalin-fixed	Fat stains Amyloid stains
(7)	Cryostat sections fresh unfixed	Oxidative enzymes Hydrolytic enzymes Almost all techniques after suitable fixation
(8)	Cryostat sections from fixed block	Hydrolytic enzymes

medium such as paraffin wax, and without having to subject the tissue to the necessary pre-treatment with organic solvents and heat implicit in normal tissue embedding. In normal histology the tissue is partly protected from the ill-effects of such treatment by the process known as fixation. For a *histochemist's* purposes however the protection offered by fixing the tissue is totally inadequate.

As we have seen, the very process of fixing the tissue, while it preserves to a certain extent the architectural integrity of the tissue, inevitably destroys many of the substances which the histochemist wishes to demonstrate. The solution to this problem lay in the development of techniques for producing thin sections from a piece of tissue which has been *frozen hard*. The intrinsic water content of the tissue, when frozen, acts as a very suitable supporting medium for the tissue and renders the tissue rigid enough for thin sections to be cut from it using specialised *microtomes* such as the freezing microtome and *cryostat*.

Although the formation of ice within the tissue when it is frozen, and the subsequent dissolution of the ice when the section is thawed, may lead to some *architectural distortion* and other artefact, the substances in which the histochemist is interested remain largely unaltered and are capable of being demonstrated by histochemical reactions.

Sections produced in this way contain the full complement of substances, and the tissue has been exposed to no damaging agent other than a low temperature; this seems to matter very little from a practical point of view. The ability to consistently produce high quality thin frozen sections, using both a freezing microtome and a cryostat, is the first requirement of a technologist involved in histochemical work; the procedures are dealt with in the relevant chapters later in this book.

It can be seen that the tissues sectioned in this way do not need the protection offered by preceding fixation. However this protective effect is not the only function of fixatives; they are also important in fixing the chemical substances in place within the cell cytoplasm and preventing diffusion (therefore facilitating accurate localisation), and also may act as mordants, enhancing the staining characteristics of certain substances and structures.

In certain circumstances fixation has an important part to play in histochemistry, although it is avoided completely if any of the oxidative enzymes are to be demonstrated. *Paradoxically*, accurate localisation of sites of hydrolytic enzyme activity is considerably enhanced if the tissue has been properly fixed; fixation reduces the amount of *diffusion* of the enzyme which inevitably occurs when the frozen section thaws.

Although the process of fixation reduces by a small amount the quantity of demonstrable hydrolytic enzyme present, this is of no consequence practically and the improved localisation makes fixation well worthwhile. The most widely used fixative for histochemical purposes is formol calcium, fixation usually being carried out at 4°C. Although fixation is valuable in the demonstration of hydrolytic enzymes, it must be for a limited time only (usually about 12 hours). The longer the tissue is kept in fixative, the more appreciable will be the loss of demonstrable enzyme activity. Two more sophisticated methods based on the freezing of tissue are *freese drying* and *freese substitution.*

In freeze drying, the tissue is frozen, and the water (in the form of ice) is completely removed in conditions of partial vacuum and at sub-zero temperatures. In freeze substitution the water in the tissue is replaced by an organic solvent, again at sub-zero temperatures, but at normal *atmospheric pressure*. These two techniques, although they require special equipment and some expertise, are most valuable in the histochemical identification and localisation of carbohydrates and proteins.

Freeze-dried tissue blocks are best fixed in formalin vapour before embedding and sectioning; in freeze substitution fixation can be effected by incorporating a fixative such as picric acid in the substituting fluid. Now that the histochemist is equipped with adequate tools for his job, the number of substances for which histochemical methods have been devised is increasing rapidly. Adequate histochemical methods now exist for the demonstration of *carbohydrates*, *proteins*, *lipids*, *nucleic acids*, *pigments*, *amines* and a very wide range of enzymes.

However, we must accept that most of the histochemical reactions which we use are embarrassingly non-specific and that we are unable to identify substances with the accuracy of the biochemist with the reagents at our disposal. At this juncture it is perhaps relevant to mention the very important part which the chemical manufacturing industry, and a handful of firms in particular, have played in the development of histochemistry.

They have continually striven to produce reagents, and particularly enzyme substrates, of such purity and specificity that there is now some hope of increasing the specificity of our histochemical reactions. Histochemists have used considerable ingenuity in devising methods to increase the specificity of their reactions, and the use of such techniques as specific enzyme inhibition has led to more accurate identification of certain enzymes. A good example of this is in the elucidation of the various types of '*esterase*'.

A very recent development in histochemistry has been the application of a limited number of histochemical methods (mainly involving the demonstration of enzymes) to blocks of tissue which are subsequently to be processed for electron microscopy. In this way we can study the ultrastructural localisation of enzyme activity within cell organelles.

Histochemistry is fast gaining *respectability*; it has long been used as a research tool in all branches of the biological sciences, and is now being increasingly used in *diagnostic histopathology*. If we wish this to continue we must ensure that any new methods are simple enough to be performed in a routine laboratory and that they are consistently reproducible.

2

FIXATION

The requirements of a fixative vary, depending upon whether the tissue is to be used for the *demonstration* of *enzymes*, or for other purposes. The main functions of a fixative are:

(*a*) to preserve the tissue
(*b*) to prevent diffusion
(*c*) to protect the tissue from subsequent treatment.

For tissue that is processed for *paraffin wax* embedding the following factors are also relevant.

(*a*) The hardening of the tissue by the fixative
(*b*) The rate of penetration of the fixative
(*c*) The conversion of *semi-liquids-gels*-to *semi-solids* (4) The effect on staining reactions.

PRESERVATION OF THE TISSUE

The first consideration of fixation must be to preserve the tissue as near as possible as it was in life, whilst at the same time not altering the chemical composition or *localisation* of the constituents of the tissue. This with present day *fixatives* is not wholly possible; however, thought must be given to what is required of the tissue before a means of fixation is chosen. *Tissue* on removal from the body, if left in the air, will dry, shrink and undergo *bacteriological* changes and autolysis (chemical reactions taking place by the enzymes within the tissue cells).

If the tissue is placed in water, the tissue swells rapidly and is unrecognisable. Upon *death*, *autolysis* and *putrefaction* (the tissue being affected by bacteria from an outside source) take place. To avoid these changes, fixation must be carried out as rapidly as possible if the tissue is to be preserved in a state *comparable* with that in life. The preserving fluid (fixative) must not cause swelling of the tissue, it must stop chemical reactions taking place with the autolytic enzymes, whilst not removing all other enzymes from the tissue. The preservative must also cause as little *shrinkage* of the tissue as possible.

Preservation by Freezing

Tissues *frozen* rapidly to temperatures of –70°C and below are well preserved. There is no major loss of enzyme activity and diffusion is prevented. The chemically reactive constituents remain *unaltered* until the tissue is thawed.

Prevention of Diffusion or Loss of Substance

For the majority of histochemical procedures, it is necessary that the substances to be demonstrated are in their true position. Despite all the fixatives to hand today, this is not always possible. In some instances, the substances are affected and moved by the fixative itself, e.g. the *diffusion* artefact shown in glycogen. It is possible to avoid this effect with glycogen by using freeze drying, followed by *formalin* vapour fixation.

In other cases, the fixative by its action on cell constituents will render the substances insoluble to other reagents and hence demonstrable by *histochemical techniques*. The correct choice of a fixative is vital. Formaldehyde does not act upon lipids, and is therefore ideal for their demonstration, whereas any of the many alcohol-containing fixatives cannot be used, because of the *extraction* of *lipid material* by the alcohol.

Protecting the Tissue from Subsequent Treatments

After suitable fixation, the tissue is generally processed either through to paraffin wax or as an alternative, it is quenched and frozen sections are cut. In both of these *procedures*, the tissue is subjected to fluids or actions (thawing) which will cause loss of substances in the first technique, or diffusion of substances in both techniques.

These can be overcome to a certain degree by the correct fixation at the start of the process for *paraffin-embedded* material, and by fixation before freezing when demonstrating *lysosomal enzymes*.

Hardening of Material for Paraffin Processing

To obtain a suitable consistency for cutting paraffin sections, the tissue must be hardened by the fixative. All the fixatives in routine use will harden the tissue, The problem is that some will overharden the material, if it is left in too long. *Hardening* also allows very soft and friable tissues to be handled without undue damage during the treatment before *embedding*.

Rate of Penetration of Fixative

Before a tissue can be considered fixed, the fixative must have fully penetrated it. The rate of *penetration* varies with different fixatives; *formaldehyde*, for instance, has a rapid rate of penetration compared with picric acid, which *penetrates* at a moderate rate. *Osmium tetroxide* has a slower rate than either.

The faster the rate of penetration of a fixative the better, as autolytic changes may still occur in the centre of a specimen despite it being in a fixative. The question of penetration is more complicated in the case of *compound fixatives*. Here, if one agent causes swelling, another usually causes shrinkage, so that uneven fixation sometimes results.

Effect of Fixation on Staining Reactions

The choice of fixative usually depends upon three factors:

(*a*) which stain or reaction is to be applied to the section;

(*b*) whether there is any urgency; and

(*c*) the size of the piece of tissue.

If the reason for processing the tissue is to demonstrate glycogen, then cold Bouin's or Gendre's would be used. However, if the *Feulgen reaction* was to be applied, Bouin's would not be used, as it causes *overhydrolysis*.

Some fixatives act as a *mordant*, a constituent of the fixative making a chemical linkage between the dye and the tissue. When this type of fixative is used, more specific staining occurs. An example of this is the use of mercury in a fixative, which gives the tissue an affinity for *trichrome stains*.

Freezing Instead of Fixing

For some histochemical procedures, it is necessary to use unfixed material. When this is required, the tissue is *rapidly frozen* on removal from the body. The freezing of the tissue preserves it by stopping the autolytic enzymes from reacting and, whilst the tissue remains frozen, it remains preserved.

On thawing, the tissue must be placed in a fixative. As well as stopping the chemical reactions, freezing 1so brings the tissue to a solid state, the ice in the tissue acting as the embedding medium when the sections are cut.

The preservation of enzymes is difficult. The optimal method for *hydrolytic enzymes* is fixation in formol calcium or formol saline at 4°C. The enzymes are less affected by the fixatives at low temperatures, and little loss of enzyme activity occurs. After this fixation, the tissue is placed in gum sucrose for 24 hours. The blocks are then blotted dry and frozen. Sections are cut in a *cryostat* and the histochemical method applied. The prefixation of the block stops the diffusion of the enzymes from their true localisation when the section is thawed before *incubating*.

GENERAL CONSIDERATIONS

The fixation of proteins also needs careful consideration, as the majority of fixatives react with protein groupings. Some fixatives precipitate proteins, e.g. alcohol, while others undergo chemical reactions with proteins, e.g. formaldehyde. Fixatives produce the best results when used at temperatures just above freezing point.

For obvious reasons, however, this is not possible in routine histology laboratories. The size of the piece of tissue to be fixed should be as small as possible (a thickness of 0·5 cm is ideal for routine use). As the fixative penetrates rapidly, the time of fixation can be kept to a minimum.

To obtain rapid fixation, the tissue should be covered by 10 to 20 times its own volume of fixative. Fixatives as a general rule tend to be acid, so for histochemical methods it is usually necessary to adjust the pH to *neutrality*.

FIXING SOLUTIONS

Formaldehyde

This is the most popular fixing agent in histology. It may be used on its own or in other compound fixatives. Formaldehyde, which is a gas, is normally available as a 40 per cent solution in water. When solutions of formalin are prepared, this 40 per cent solution must be taken as 100 per cent, e.g. to prepare a 10 per cent solution of *formalin*, 10 parts of 40 per cent formaldehyde are added to 90 parts of water. Solutions of formalin are *invariably acid*, due to the formation of small amounts of *formic acid* in the *formaldehyde*.

For all histochemical, and the majority of *histological*, techniques it is necessary to bring the pH of formalin to neutral. This can be done by employing either a buffer (sodium phosphate), or by placing calcium carbonate at the bottom of the container holding the *formaldehyde* solution. Fixation in acid formalin over a lengthy period produces formalin pigment, especially in blood-containing tissues. If the tissue

Table 2.1: Aldehyde Fixatives.

Aldehyde	*%*	*Buffer*	*Final pH*
Acrolein	10		6.8-7.2
Glutaraldehyde	6		7.2
Crotonaldehyde	10		7.4
Formaldehyde	10	Phosphate buffer	7.2-7.4
Hydroxyadipaldehyde	12		7.4
Acetaldehyde	10		7.2-7.4
Glyoxal	4		6.8-7.2

has been fixed in acid formalin over a long period of time, difficulty may be encountered in making acid dyes such as eosin stain correctly.

Formaldehyde will not fix lipids, nor will it remove them, which allows for their demonstration. The following formaldehyde solutions are amongst those used in histochemistry and histology:

(*a*) 10 per cent *formol saline*

(*b*) 10 per cent *formalin* (*aqueous*)

(*c*) 10 per cent *neutral formalin*

(*d*) 10 per cent *formol calcium*

(*e*) 10 per cent *formalin in alcohol*

(*f*) 10 per cent *formol sucrose*

(*g*) Formaldehyde vapour.

As in routine *histology*, formaldehyde is probably the most used fixative in histochemistry. It is an ideal fixative for tissue in which lipids are to be demonstrated, as lipids are chemically unaltered by formaldehyde fixation. *Phospholipids* are well preserved by formaldehyde fixation when calcium has been added to the fixative.

In enzyme histochemistry, the majority of hydrolytic enzymes are well preserved with fixation in formol calcium at 4°C. The amount of enzyme activity is reduced if the fixation time is extended or if it is carried out at room temperature. Formaldehyde vapour is recommended for the fixation of freeze dried tissues and produces excellent demonstration of mucosubstances including *glycogen* and *proteins*. Nucleic acids are also well preserved after vapour fixation.

Other Aldehyde Fixatives

Other aldehydes, besides formaldehyde, may be used as fixatives. They are especially useful when enzyme histochemistry is followed by *electron microscopy*. Sabatini, Bensch and Barrnett produced a paper dealing with nine different aldehydes. They buffered the fixatives to between pH 5.5 and 7.6.

The hydrolytic enzymes worked satisfactorily after most of the fixatives, as did NADH and NADPH diaphorases. Flitney, using seven different aldehydes, investigated the rate of fixation of these fixatives with *albumen*, and their effect on enzyme activity in cryostat sections. In regard to the fixation of *cryostat* sections, he found that aldehydes which fixed the albumen most rapidly also destroyed most enzyme activity.

This could be overcome, however, by using these fast acting aldehyde fixatives for very short times, i.e. for an 8 micron cryostat section, 30 seconds to 1 minute was usually sufficient. The two aldehydes that gave good fixation with reasonable *morphological* preservation and with only a small amount of enzyme loss were *glutaraldehyde* and *acrolein* when used for 60 seconds. These aldehydes have not gained popularity as block fixatives because of the time taken to penetrate to the centre of the tissue. The advantage to be gained by short fixation is lost.

Alcohol

This is rarely used on its own in histology because of the damage it causes by shrinkage and excessive hardening of the block. It is, however, used more often in histochemistry, especially for fixation of cryostat sections, to demonstrate enzymes. Alcohol is a good fixative as the enzymes are almost unaffected by cold alcohol (4°C) with the exception of the esterases.

For the demonstration of glycogen, alcohol is occasionally used as an 80 per cent solution. This, while being an ideal fixative for glycogen, has drawbacks from a *morphological* point of view. When used in compound fixatives, however, alcohol will give acceptable results. It will precipitate proteins, and will remove lipids. It also makes the freezing of tissue and the subsequent sectioning difficult. The following are the more common fixatives which contain *alcohol*.

(*a*) Formol alcohol
(*b*) Carnoy
(*c*) Clarke's
(*d*) Gendre's fluid
(*e*) Acetic alcohol formalin (AAF)
(*f*) Wolman's fixative.

(The formulae for the above fixatives can be found at the end of this chapter.)

Table 2.2 : Suggested Fixatives for Histochemical Methods.

Methods	*Fixatives*	*Temp*	*Block time (h)*	*Section time (min)*
Acid and alkaline phosphates	Acetone 100%	4	18	30-60
	90% Alcohol	4	24	30
	Formol calcium	4	24	30
	10% Formol saline	4	24	30
Non-specific esterase	Acetone 100%	4	18-24	30-60
E600 Resistant esterase	Formol calcium	4	24	30-60
	10% Formol saline	4	18-24	30-60
Glycogen	Gendre or Bouin	4	18-24	
	Formol vapour after freeze drying			
Carbohydrate methods generally	10% Formol saline	22	2-24	30
	Formol alcohol	22	6-24	10-20
	90% Alcohol	22	6-24	30
	Formol vapour after freeze drying			
All types of lipids	10% Formol calcium	22	18-24	-
	10% Formol saline	22	18-24	-
	10% Neutral buffered solution	22	18-24	-
Protein	Formol vapour after freeze drying	60	2	30
Methods	10% Formol saline	22	18-24	30-60
	Carnoy	22	1	1-5
	Formol alcohol	22	6-24	5
Acridine orange technique	Carnoy	22	1	1-5
	100% Alcohol	6-24	30-60	
Nucleic acid methods	Carnoy	22	1	1-5
	10% Formol saline	22	18-24	30-60
Pigment methods	10 % Neutral buffered formol saline	22	18-24	30-60
Calcium	10% Neutral buffered formol saline	22	18-24	30-60

Osmium Tetroxide

Osmic acid finds little use in routine *histology* and *histochemistry*, with the exception of the demonstration of some lipid-containing structures, although it is in constant use in electron microscopy where it is used in a I per cent buffered solution. The fixation of moderate-sized pieces of tissue in osmium produces large amounts of shrinkage and uneven penetration of the tissue.

It can be used to render lipids insoluble. It can also be used as a *vapour fixative* with *freeze* dried material. This fixative is very damaging to enzymes.

Acetone

This reagent may be used in histochemistry as a fixative of cryostat sections. The acetone is used at 0-4°C. It has frequently been used as a fixative for pieces of tissue to be processed to paraffin wax, and hydrolytic enzyme methods applied to the sections. *Acetone* is a fairly rapid fixative but does, in the author's experience, cause shrinkage of the tissue.

It is frequently employed as the first step in the freeze substitution technique. At – 70°C it will not remove much lipid material from the tissue, nor will it fix the tissue, whereas it will be substituted for the ice in the tissue. It is rarely used in any compound fixatives, certainly not in any of the standard fixatives. At 4°C, it will fix cryostat sections (5-20 microns (μ m)) in 1 hour or less and will fix small pieces of tissue overnight at 4°C.

Picric Acid

This gives good fixation of tissue blocks. It is a rapid fixative that hardens the block well, without causing much shrinkage. Picric acid is used in many compound fixatives, of which Bouin's and Gendre's are the most popular. Any fixatives containing this acid are recommended for the demonstration of glycogen and histological trichrome stains.

Tissue fixed in picric acid must be washed in either alcohol or water until the yellow colour of the block is removed. Lillie states that it is more rapid to process, embed and section in the normal way and to remove the yellow colour from the section before staining. *Picric acid* precipitates proteins and combines with some of them. In some instances, these are soluble in water, but they may be treated with alcohol first and then they become insoluble.

Mercury-Containing Fixatives

These fixatives are little used in histochemistry. *Mercuric chloride* has a rapid but uneven penetration and only thin blocks of tissue should be fixed. It produces a poor result with *glycogen*. If the tissue is overfixed, the blocks become hard and sections are difficult to produce. *Mercuric chloride* also causes a great deal of *shrinkage* and is therefore rarely used alone.

It is usually employed in a compound fixative with *formaldehyde* or *acetic acid*. All tissue fixed in mercuric-containing fixatives will have a mercury precipitate, which must be removed by treatment with dilute iodine, followed by immersion in 3 per cent *sodium thiosulphate*.

FIXATIVES FOR USE IN HISTOCHEMISTRY

Acetone

Absolute acetone at 4°C

Alcohol

Alcohol 80-100 per cent, at 4°C

10 per cent Formol saline

Formaldehyde 40 per cent	100 ml
Sodium chloride	9 g
Distilled water	900 ml

10 per cent Formalin

Formaldehyde 40 per cent	10 ml
Distilled water	90 ml

10 per cent Neutral buffered formalin

Formaldehyde 40 per cent	10 ml
Distilled water	90 ml
Sodium dihydrogen phosphate (anhydrous)	350 mg
Disodium hydrogen phosphate (anhydrous)	650 mg

10 per cent Neutral formalin

Formaldehyde 40 per cent	10 ml
Distilled water	90 ml
Calcium carbonate chips to cover bottom of container	

10 per cent Formol calcium

Formaldehyde 40 per cent	10 ml
Distilled water	90 ml

Calcium chloride	1,100 mg (or until pH reaches 7·0)

Formol sucrose

Formaldehyde 40 per cent	10 ml
Sucrose	7·5 g
0·2M Phosphate buffer, pH 7·4	90 ml

Carnoy fixative

Ethyl alcohol	60 ml
Chloroform	30 ml
Glacial acetic acid	10 ml

Clarke's fixative

Absolute alcohol	75 ml
Glacial acetic acid	25 ml

Formol alcohol

Formaldehyde 40 per cent	10 ml
Absolute alcohol	80 ml
Distilled water	10 ml

Acetic alcohol formaldehyde (AAF)

Formaldehyde 40 per cent	10 ml
Glacial acetic acid	5 ml
Absolute alcohol	85 ml

Gendre's fixative

Picric acid (saturated in 95 per cent alcohol)	85 ml
Formaldehyde 40 per cent	10 ml
Glacial acetic acid	5 ml

Bouin's fixative

Picric acid (saturated aqueous)	75 ml
Formaldehyde 40 per cent	25 ml
Glacial acetic acid	5 ml

Formol sublimate

Mercuric chloride (saturated aqueous)	90 ml
Formaldehyde 40 per cent	10 ml

Flemming's fixative

1 per cent Chromic acid	60 ml

2 per cent Osmium tetroxide	16 ml
Glacial acetic acid	4 ml

Newcomer's fixative

Isopropanol 50 ml Propionic acid	25 ml
Petroleum ether	8.3 ml
Acetone 8.3 ml Dioxane	8.3 ml

Heidenhain's 'Susa.'

Distilled water	76 ml
Mercuric chloride	4.5 g
Sodium chloride	500 mg
Trichloracetic acid	2 g
Acetic acid	4 ml
Formaldehyde 40 per cent	20 ml

Zenker's fixative

Distilled water	95 ml
Mercuric chloride	5 g
Potassium dichromate	2.5 g
Sodium sulphate	1 g
Glacial acetic acid, added before use	5 ml

Helly's fixative

Distilled water	95 ml
Mercuric chloride	5 g
Potassium dichromate	2.5 g
Sodium sulphate	1 g
Formaldehyde 40 per cent is added before use	5 ml

6 per cent glutaraldehyde

Glutaraldehyde (25 per cent)	24 ml
0.1M Phosphate buffer, pH 7.4	76 ml

On storage, the glutaraldehyde will become acid, pH 2.5 to 3.0. The final pH of the above fixative should be checked and adjusted to 7.0 to 7.2 if necessary.

1 per cent acrolein

Acrolein	1 ml
0.1 m Phosphate buffer, pH 7.4	99 ml

The final pH of this fixative should be checked and adjusted to 7.2 to 7.6 if necessary.

3

Paraffin Sections

Routine paraffin sections have been produced in hospital and other laboratories for many years. In order to produce these sections in such a manner that they can be of use a number of criteria must be satisfied. The tissue in section must resemble as closely as possible its appearance in life. The most important factor is to '*fix*' or preserve the tissue. This fixation also hardens the tissue, but not enough to allow thin sections to be produced from the block.

To facilitate the production of these sections, the fixed tissue has to be embedded in a medium to render it firm and rigid. This is *traditionally* carried out by placing the tissue in paraffin wax or, for specialised procedures, in suitable resins. Few of the *embedding* media available are *miscible* with the water invariably present in the tissue and fixative.

This gap is bridged by the process known as *dehydration* and *clearing*, the removal of water from the tissue blocks being attained by treatment with increasing concentrations of alcohol until no water remains. As the embedding media are not alcohol-miscible an organic solvent must be used to remove the absolute alcohol; this is the process known as '*clearing*'. A number of organic solvents may be used of which *chloroform*, *toluene* and *xylene* are examples.

Following the removal of the alcohol, the tissues are impregnated in molten paraffin wax at 58°C. When the wax cools down to room temperature it sets as a hard block in which the tissue is contained. The support offered by the wax enables thin sections of the tissue to be cut on a *microtome*. These sections are ideal for treatment with the many *dyes* and *stains* available to demonstrate the histological features of the tissue. The use of these paraffin sections in histochemistry is somewhat restricted. They can be used for the demonstration of *non-enzymatic components* and under carefully controlled conditions certain *hydrolytic enzymes* may be demonstrated but in a considerably reduced amount. The effects of routine paraffin processing on a piece of tissue are predictable. They are listed in Table elsewhere in this chapter.

It will be observed that very little enzyme activity will remain; it is possible, however, for very small amounts of *hydrolytic enzymes* such as alkaline phosphates to be demonstrated. The accuracy of localisation is open to doubt. The *alteration* of standard processing methods will allow the demonstration of more enzyme activity.

While these methods seem to cause much morphological damage (e.g. shrinkage) they are well worth trying if no other means of processing the tissue are available.

Table 3.1 : Effects of Routine Processing Methods on Tissue for Histochemical Demonstration

Fixation at room temperature Most fixatives	Loss of enzyme activity Denatured proteins (depending upon fixative and length of fixation) General diffusion of enzymes Loss of some carbohydrates Diffusion of other carbohydrates
Dehydration in alcohol (room temperature)	Loss of enzyme activity General diffusion of enzymes Loss of some lipids Shrinking, hardening
Clearing in either xylene, chloroform, toluene, etc.	Loss of enzyme activity Loss of lipids Shrinkage, hardening (depending upon clearing agent)
Wax embedding Paraffin wax, 56-60°C	Loss of enzyme activity Loss of lipids Shrinkage and hardening

The reagents that can be used to *dehydrate* and clear tissue for normal histological purposes are numerous. The fluids used when processing tissue for histochemistry should be as inert as possible. This causes the smallest loss of enzyme activity. If these solutions, e.g. acetone and petroleum spirit, are used at low temperatures (4°C), the retention of the enzymes is further increased. Therefore fixation, dehydration and most of the clearing stages are carried out at 4°C. For the full method see Method 3.1.

As well as loss of *enzymes* with clearing agents, heat also causes a big loss in enzyme activity, especially above 40°C. It is advisable to use a paraffin wax with a melting point as low as can be *conveniently* handled, usually between 45°C and 54°C. The author made a comparative study of four different processing methods, and a summary of the results is given in Table elsewhere in this chpater.

It can be seen from the results that while no enzyme activity remained in the routine processed material, the amount in the specially processed block and the freeze dried tissue were very similar, demonstrating that the loss of enzyme activity was probably due to the paraffin wax and the heat, rather than the *dehydrating solutions* used.

Paraffin sections may be used for the demonstration of the following groups of substances: *proteins*; *carbohydrates*; *nucleoproteins*; *coloured pigments*, and *phospholipids*.

Proteins

It is possible to demonstrate most structural proteins in routine processed *paraffin material*, although freeze drying is the ideal method of processing. Fixation in formalin or *acetic ethanol* is recommended and *dehydration* and *clearing* times should be kept to a minimum.

Nucleoproteins

These may be demonstrated very successfully in routine paraffin material, Carnoy's fixative possibly being the best for this purpose.

Carbohydrates

This is a complex group of substances for which there are many different staining methods. Most of these methods will work quite *satisfactorily* on routine paraffin processed material, although superior results can be obtained using the freeze *drying technique*.

Pigments

The majority of methods used to demonstrate pigments can be carried out with routine processed paraffin material.

Lipids

These cannot be demonstrated in paraffin sections apart from phospholipids, but even with *phospholipids* it is preferable to use frozen sections.

Naturally Occurring Fluorescence

Natural fluorescence is demonstrated most efficiently in freeze dried formol vapour fixed material.

Method 3.1

Paraffin processing at 4°C

(*a*) Select tissue block and trim to 1.0 × 1.0 × 0.5 cm or smaller

(*b*) Place in precooled acetone at 4°C overnight

(*c*) Transfer to petroleum ether at 4°C for 1 h

(*d*) Transfer to petroleum ether at 4°C for 12h

(*e*) Transfer to petroleum ether at 20°C for 1 h

(*f*) Place in 40°C paraffin wax at 42°C for 15 min

(*g*) Place in 40°C paraffin wax at 42°C for 15 min

(*h*) Embed in 40°-45°C wax

This method allows the demonstration of some of the hydrolytic enzymes, but does cause shrinkage of the block.

Method 3.2

Paraffin processing double embedding at 4°C

(*a*) Select and trim block to 1 × 0.5 cm or smaller

(*b*) Place blocks into precooled acetone at 4°C for 18 h

(*c*) Transfer to fresh acetone at 4°C for 6 h

(*d*) Dehydrate in acetone at 4°C for 12 h

(*e*) Dehydrate in acetone at 4°C for 12 h

(*f*) Transfer blocks to 2 per cent celloidin at 4°C for 12 h

(*g*) Transfer blocks to 2 per cent celloidin at 4°C for 12 h

(*h*) Blot blocks to remove excess celloidin

(*i*) Place blocks into chloroform (1) at 20°C for 1 h

(*j*) Place blocks into chloroform (2) at 20°C for 2 h

(*k*) Embed in paraffin wax.

The method, although lengthy, does allow the demonstration of some hydrolytic enzymes. It causes less shrinkage artefact than Method 3.1.

4

FROZEN SECTIONS

The standard histological processing procedure, although capable of producing excellent sections suitable for many purposes, does have a number of drawbacks and the majority of these affect the histochemist. As far as the *histochemist* is concerned, routine paraffin processing alters the chemical reactions within the tissue. It will probably cause complete loss of *demonstrable enzymes*, the loss of other soluble substances (e.g. lipids), or the *alteration* and *diffusion* of other substances. For these reasons, and those listed below, it is necessary to be able to produce sections that have not been through this technique.

If the tissue is frozen it becomes of a consistency that will allow sections to be cut from it. The water in the tissue, on turning to ice, gives the tissue sufficient support to allow this *sectioning* to take place. In *histochemistry* there are, applications where fixed tissue sections are preferred to unfixed frozen sections, for example in the *demonstration* of lipids and some *hydrolytic* enzymes; the reasons for this are discussed in the relevant chapters.

The fixed tissue, with the additional water from the fixative, can also be cut frozen, and is considerably easier to handle on a freezing microtome than on a *cryostat*. Frozen sections may be cut from fresh unfixed tissue and from fixed tissue (usually formalin fixed). They may be prepared using one of the following techniques.

(*a*) Conventional freezing microtome

(*b*) Standard microtomes with thermomodules attachments

(*c*) Cryostat.

The last technique is fully dealt with in other chapter of this book.

USES OF FROZEN SECTIONS

Fresh Unfixed Frozen Sections

These sections, which can prove very difficult to handle once they are cut, are little used today if a *cryostat* is available. Their main uses were and are as follows.

(*a*) *Urgent biopsy* specimens from the *operating theatre*

(*b*) Fat stains (not recommended)

(*c*) Histochemical methods, requiring unfixed sections, e.g. usually oxidative enzymes (dehydrogenases), glucose-6-phosphatase and others)

(*d*) Sections for quantitative work

(*e*) Immunofluorescence techniques

(*f*) Autoradiography.

Fixed Frozen Sections

These sections, which are used in many laboratories, are considerably easier to handle throughout the histochemical method than the unfixed material. Their main uses are as follows.

(*a*) Urgent biopsy specimens from theatre

(*b*) Fat stains

(*c*) Histochemical methods requiring fixed free-floating frozen sections, e.g. accurate localisation of some hydrolytic enzymes.

(*d*) Fluorescent techniques

(*e*) Demonstration of soluble substances

(*f*) Autoradiography

(*g*) Impregnation techniques (silver methods).

The method of cutting urgent biopsy specimens is given in Method 4.2 at the end of this chapter.

USE OF FREEZING MICROTOME

Cutting of Unfixed Frozen Sections

This technique involves freezing a piece of fresh tissue, 2 cm × 2 cm × 0.75 cm or smaller, onto the stage of a freezing microtome. This is accomplished by placing a drop of water on the stage and holding the tissue firmly against it with the *index finger*, while giving repeated blasts of carbon dioxide. The water turns to ice, thus holding the tissue firmly onto the stage. Further *blasts* of *carbon dioxide* are given until the block is completely *frozen*. This can be observed by the colour change of the tissue from its natural colour to a glossy white.

Once the tissue is frozen it is necessary to cool the knife. In freezing *microtomes* not equipped for *knifecooling* with carbon dioxide gas, the knife is cooled with pieces of solid carbon dioxide.

When knife and block have both been cooled, the block must be brought up to the correct cutting temperature. This is done by overcooling the tissue-and then warming the surface of the block by placing a finger on it. The knife is then passed over the block. If the tissue is too cold, the sections will *crumble* and *fragment*. The block should then be warmed again, and the procedure repeated until satisfactory sections are cut. The sections should curl and remain on the knife.

They are removed by stroking the finger towards the knife edge from the back of the knife, or by using a brush. The sections are placed in saline and are then ready to be mounted on slides or coverslips. Alternatively, they can be stained free-floating, as described later.

Cutting of Fixed Frozen Sections

Fixed tissue is easier to cut on a *freezing microtome* than unfixed tissue but in a cryostat the situation is reversed. *Formaldehyde* is a good fixative for fixed frozen sections : alcohol-containing solutions should be avoided because of their low freezing point, and dichromate-containing fixatives make the tissue brittle and more difficult to cut.

It is not necessary for the knife to be cooled for cutting fixed tissues, for when the sections are cut they will not form a '*slush*' against the knife. The fixed tissue blocks need a warmer cutting temperature because of the increased water content of the tissue; the sections are easier to handle after cutting than are unfixed sections. Otherwise the technique is as for unfixed blocks.

Limiting Factors Concerning Freezing Microtome Technique

A few of the many limiting factors of this technique are listed below.

(*a*) As a rule sections 10 microns or thicker only can be cut
(*b*) Only small thin blocks can be used
(*c*) Certain tissues are difficult, if not impossible, to cut
(*d*) Freezing artefact is likely to be present in some tissues
(*e*) Tissues must be fresh (if unfixed sections are required)
(*f*) Correct cutting temperature is difficult to maintain
(*g*) Lack of supporting medium causes shattering of fine structures.

Embedding Tissues in Gelatin

If tissues are friable, or if many small fragments are assembled to make one block, it is *advisable* to embed the tissue in *gelatin* before cutting frozen sections. One of the disadvantages of this technique is that the gelatin is frequently also stained. The technique for gelatin embedding of tissue is given at the end of this chapter in Method 4.1.

Fixation of Blocks

The fixation of blocks of tissue for the frozen section technique depends upon the *histochemical* method to be carried out on the section. However, as routine fixatives, formol saline or formol calcium are recommended. It is advisable to avoid fixatives with an *alcohol* constituent because of the inhibitory effect of the alcohol on freezing. Fixatives causing excessive *hardening* of the tissue are also to be avoided.

Picking Up Frozen Sections on Slides

Many of the techniques using frozen sections call for the *staining* of the sections *free-floating*. The *mounting* of the section comes at the end of the process. Where possible, it is advisable to pick up the sections on slides before staining, the exception being in the demonstration of lipids by the *Sudan methods*.

Mounting Frozen Sections on Slides

To pick up fixed frozen sections, coat a *microscope* slide with *glycerin* albumen and float the section onto *distilled water*. If the section is flat and *crease-free*, place the slide underneath the section and bring it upwards until it touches one edge of the section. At this point, bring the slide to the *vertical position* and *slowly raise* it out of the water.

If the section is creased, allow the excess water to drain away, then replace the slide in the water until half the section is free-floating; the creases are then removed and the slide withdrawn from the water. Should the other half of the section contain creases, the slide is reversed and the exercise repeated. The sections are then drained, before being blotted and allowed to dry.

If the section is badly creased and folded when it is initially placed on the water, it can be floated onto 70 per cent alcohol for 30 seconds and then returned to the distilled water. The different surface tensions in the two solutions cause the section to spin, coming eventually to rest in a crease-free state on the top of the water.

The sections may be dried at room temperature or at 37°C. As an alternative to using *glycerin albumen* to retain frozen sections on the slide, the following technique is suggested.

The section is picked up on a *clean slide* and *blotted* firmly with filter paper soaked in absolute alcohol; it is then washed in absolute alcohol. Repeat the blotting, then coat the slide in 0.5 per cent *celloidin* and allow to *partially dry*.

The slide is now ready to be washed in water and to be stained. At the end of the staining technique, during the *dehydration*, the slide is placed in absolute alcohol/ether mixture 1:1 to remove the *celloidin* before being placed in *xylol*. This technique is unsuitable for any method demonstrating soluble substances (e.g. *lipids*), or impregnation methods (e.g. *silver*).

Staining Free-Floating Sections

This is the method of choice when carrying out staining procedures such as oil red 0, or the silver *impregnation* techniques. The sections, preferably fixed frozen sections, are placed into distilled water after cutting. The *equipment* needed for this method consists of embryo staining dishes (the same number as there are steps in the staining technique), a glass rod shaped like a hockey stick and a *dissecting needle*.

The sections are placed in the first staining dish by manipulating the section out of the distilled water by means of the '*hockey stick*'. It is possible to wrap the sections around the hockey stick without causing them any *damage*. A section is passed along the staining dishes in this manner until it reaches the *distilled water* at the end of the procedure. From the distilled water it is picked up on an *albumenised slide*, as described earlier, with the aid of the hockey stick or a dissecting needle.

USE OF THERMOMODULE ATTACHMENT

In recent years the technique of using thermomodules to maintain the block temperature, and in some cases the knife temperature, has been gaining *popularity*. The first report of this technique was by loffe and further papers were published by *Brown* and *Dilly* and *Hardy* and *Rutherford*. All these authors stated that frozen sections could be cut with ease. *Brown* and *Dilly* listed the following five advantages that this system had over the conventional carbon dioxide technique.

(1) Convenience: The need for carbon dioxide gas is eliminated
(2) Accurate temperature control
(3) Optimal cutting temperature can be maintained indefinitely
(4) Serial sections can be cut at 5 microns
(5) Adaptability of unit for other cooling uses.

The first two papers mentioned above describe the use of the modules as a means of maintaining the block temperature. *Hardy* and *Rutherford* in a further paper, described the cooling of the knife, as well as the block, to allow for cutting of all types of material. In a later paper, the authors described the use of different micro tomes with the *thermoelectric* equipment.

Using Thermomodules for Frozen Sections

Thermoelectric modules have adaptors which allow them to be fixed onto all standard *microtome stages*. The modules and their electrical equipment are produced commercially. As the thermomodules require water for the hot stage, it is necessary to place them on a bench near to a supply of running cold water.

When operated under optimal conditions, the module is capable of reaching just below -30°C, and this temperature is controllable to +1°C. To cut unfixed material, it is necessary to use the thermomodules on the knife, which can then be cooled to -20°C or below, if required.

Cutting a Fixed Block of Tissue

A piece of fixed tissue is placed onto the microtome stage together with a drop of water. The current is turned to maximum and the block will be seen to freeze. When the block is completely frozen, the *electrical* supply is reduced until a satisfactory consistency of the block is obtained. It is possible to maintain the block at this temperature indefinitely.

The cutting temperature varies with the nature of the tissue, as with all frozen section techniques. The sections are floated-out onto water or formol saline.

Cutting an Unfixed Block of Tissue

The thermomodules are attached to the knife to cool it a few degrees below the temperature of the block. Once the knife and block are at the correct temperature, it is far easier to obtain *satisfactory* sections of unfixed material with this method than by any other *technique*, apart from the *cryostat*.

If enzyme histochemical methods are to be applied to sections produced by this technique, it is necessary to freeze the tissue by one of the procedures listed in other chapter of this book since *thermomodule* freezing is often not rapid enough.

Otherwise the tissue is frozen by placing it on the thermomodule in the manner described above. Using this technique it is possible to obtain serial sections with a thickness of less than 10 microns.

Method 4.1

Gelatin embedding for friable tissues

(*a*) Fix block in 10 per cent formol saline overnight
(*b*) Wash out formalin by washing in running tap water for 6-8 h
(*c*) Place block into glycerin-gelatin mixture at 37°C for 6 h
(*d*) Transfer block to fresh mixture and embed in suitable mould and allow to cool
(*e*) Place the block of gelatin in 10 per cent formol saline overnight
(*f*) Trim away the excess gelatin and reimmerse in 10 per cent formol saline for 8 h or until required.

Gelatin	16 g
Glycerin	15 ml
Distilled water	70 ml
Thymol	1 crystal

Method 4.2

Cutting of frozen sections of urgent biopsy specimens

(*a*) On notification of an urgent biopsy, place 100 ml of 10 percent formol saline to heat on a Bunsen burner
(*b*) On arrival of biopsy remove the formol saline from the boil
(*c*) Select block to be cut
(*d*) Place in hot 10 per cent formol saline for 60-90 s
(*e*) Wash rapidly in running tap water for 10 s
(*f*) Place drop of water on microtome stage
(*g*) Transfer tissue to the microtome stage
(*h*) Freeze the tissue with rapid bursts of CO_2
(*i*) Warm the block with finger to correct cutting temperature
(*j*) Cut 6-12 sections at 10-15 microns
(*k*) Place sections in distilled water
(*l*) Pick up on glass slides as indicated
(*m*) Blot dry with filter paper
(*n*) Pass rapidly over Bunsen flame
(*o*) Rinse in water
(*p*) Stain in Harris' haematoxylin for 60 s
(*q*) Wash off stain
(*r*) Differentiate in 1 per cent acid alcohol for 10 s
(*s*) Place in Scott's tap water for 20 s

(*t*) Wash in tap water for 5 s
(*u*) Place in I per cent eosin for 5 s
(*v*) Wash in tap water for 10 s
(*w*) Dehydrate through graded alcohols to:
(*x*) Xylene
(*y*) Mount in DPX.

5

CRYOSTATS

The advent of the cryostat has contributed much to the rapid development of histochemistry since the early nineteen-fifties. Before the manufacture of the cryostat it was difficult, if not impossible, to consistently obtain thin sections of unfixed material with reasonable preservation and quality.

DEVELOPMENT OF CRYOSTATS

The first cryostat described in the literature was produced by Linderstrom-Lang and Morgensen who developed their cryostat for quantitative *cytochemical work*. Alternate sections of equal thickness were taken for biochemical estimations and for histology. The cabinet for the cryostat was kept at approximately -2D°C; this was done by means of large blocks of *dry ice* placed at the back of the cabinet. A fan was used to *circulate* the cold air from above the dry ice around the chamber.

The fan was not in operation, however, while sections were being cut. A rotary microtome was used in the well insulated chamber. Another feature of this cryostat was a device which, when pressed against the knife edge, *allowed flat*, *crease-free* sections to be cut. This device was called an anti-roll plate, or *guide-plate*. The guide-plate assembly, though much modified from the original form, is still of great importance in the operation of the *cryostat*.

The anti-roll plate is dealt with in detail later. Access to this early *microtome* was through a hinged sloping door on the front of the cabinet. In this door were two large gloved *portholes* through which the microtome was operated, a small stoppered port for the removal of sections, and also a window for observation. To combat one of the problems of rotary microtomes in cryostats, the *microtome* had a small electric heater to keep it a few degrees above freezing point. From this cryostat it was only a short step to the model produced by *Coons*, *Leduc* and *Kaplan*.

Coons-Leduc-Kaplan Cryostat

The cryostat designed by these workers was based on the earlier model, except that the cold environment was produced by refrigeration coils running round the cabinet. The latter, which was of 6 ft. capacity, was made of *stainless-steel*. The operating temperature obtained with this model was between -16°C and -18°C. The cryostat was later produced *commercially* by the Harris Refrigeration Co. of Cambridge, Massachusetts. Later models produced by this company have operating temperatures down to -40°C. An alternative Coons-type of cryostat is produced in Germany.

Dittes Rotary Cryostat

In this model a large Jung rotary microtome is used. The *actuating wheel* of the microtome is placed outside the cabinet, so that while the operator has his hands inside the chamber, the machine is operated by an assistant. To avoid rusting of the microtome, the *Dittes cabinet* is also modified to reduce the *humidity*. Later models are available with a *foot-controlled* electric motor to drive the large microtome.

Development of British Cryostats

Following the arrival in Britain of the original *Coons-type* of cryostat and its success, there was a rapid increase in the use of cryostat-sectioned material. *Coons* and his *colleagues* had designed their instrument to provide sections for use with the *fluorescent* antibody technique.

The production of frozen sections for the *demonstration* of enzymes became a major function of the equipment, as well as the increasing use for urgent *biopsy* specimens.

In 1954 Pearse started to build the prototype of the Pearse—*Slee cryostat*. In building his cryostat, Pearse intended that seven criteria be satisfied; they were:

(1) remote control

(2) instant availability for diagnostic biopsies

(3) thermostatic control between -5°C and -20°C

(4) minimum temperature below -20°C

(5) low cost and constant running

(6) automatic delivery of sections into incubating media or fixatives, and

(7) ability to produce serial sections.

The cabinet of this *cryostat* was cooled by refrigeration coils running along two of the interior walls. It was capable of reaching –30°C and an accurate *thermostat* allowed control of temperature between –5°C and –30°C.

The microtome in the original Pearse-Slee, and also in the first Bright cryostat, was a *Cambridge rocking microtome*; the controls (i.e. coarse advance, thickness setting, operating handle and anti-roll plate) were all external and could be operated through the sides or top of the chamber. The anti-roll plate was attached to the microtome and not, as in the Coons cryostat, to the *knife*. The front of the cryostat had a porthole for the removal of sections.

Current Models

The *development* of the cryostat over the last twenty years has brought many *improvements*, as well as a *diversity* of choice. The largest single improvement has been the gradual change from the rocking microtome to the *rotary*. This has allowed larger and more difficult blocks to be cut, and thinner sections to be produced.

Another welcome advance is the heated and hinged window which has entirely eliminated the early problems associated with the *frosting* of the window. The anti-roll plate, the proper adjustment of which is a major part of successful cryostat technique, has also become a more sophisticated piece of equipment with accurate *microadjustment*.

Motorised cryostats have become popular in recent years; the obvious application for this type of cryostat is in the cutting of serial sections, and in some instances where the correct cutting speed is critical to success in producing the section.

There are a number of cryostats produced *commercially* which are operated from above, i.e. the top is hinged and is removed when the microtome is in use. This type of cryostat (commonly manufactured in America), of which the Harris International, The Lab Tek and the American Optical Co. cryostats are representatives, is operated under the same conditions as other models.

Leitz have developed a very *sophisticated* cryostat known as the Histokryotom. In this model the sections are *automatically* picked up from the knife edge onto slides contained within the cryostat. Many of the functions of this cryostat can be controlled *electronically*.

Thermoelectric Cryostats

Thornburg and Mengers in their paper '*Analysis of Frozen Section Techniques*', made the observation that if the environmental temperature was between +5°C and –5°C serial sections of 2μ m could be obtained. The advent of the use of thermomodules in frozen section techniques by *Ioffe*, *Brown and Dilly*, *Rutherford*, *Hardy* and *Isherwood*, and other workers, led to the possibility of a thermoelectric cryostat being produced.

Table 5.1 : Temperature Optima for the Frigistorised Cryostat (Unfixed Rat Tissue).

Tissue	*Knife temperature (°C)*	*Block temperature (°C)*	*Environmental temperature (°C)*
Liver	–18	–10	–5
Kidney	–15	–8	–5
Skin		–35	–10
–5			
Keloid	–35	–12	–5
Patella	–40	–18	–5
Brain	–18	–15	–5
Thyroid	–20	–10	–5

The development of this machine was described by Pearse and Bancroft. The results of these experiments confirmed the findings of *Thornburg* and *Mengers* that, if it is possible to vary the temperature of both the knife and block, –20°C or below is too low a figure for the environmental temperature. In the *thermoelectric cryostat* it is possible to vary the block, knife and *chamber* temperature *accurately* for the first time and when these temperature variables were *studied*, the following facts emerged.

(*a*) Each tissue has its own optimal block (cutting) temperature

(*b*) Each tissue has its own optimal knife temperature

(*c*) The chamber temperature is sufficiently cold if maintained at –5°C.

Tables elsewhere in this chapter, reproduced from the Pearse and Bancroft paper, give the temperature optima for cutting of rat tissue, both unfixed and fixed. *Unfixed* and *fixed* tissues need different handling; the treatments are discussed separately.

CRYOSTAT TECHNIQUES

The production of thin, artefact-free sections in the cryostat is dependent upon a number of factors. These are:

(*a*) satisfactory preparation of tissue

(*b*) cryostat temperature (involving both tissue block and knife)

(*c*) microtome

(*d*) adjustment of anti-roll plate.

All these factors must be correct before successful sections can be consistently produced.

Table 5.2 : Temperature optima for the frigistorised cryostat (fixed rat tissue).

Tissue	*Knife temperature (°C)*	*Block temperature (°C)*	*Environmental temperature (°C)*
Liver	–15	–8	–5
Kidney	–10	–8	–5
Skin	–35	–5	–5
Keloid	–35	–5	–5
Patella	–40	–20	–5
Brain	–15	–8	–5
Thyroid	–25	–12	–5

Preparation of Material

Tissues to be cut in a cryostat can be either fresh frozen, i.e. unfixed, or fixed (usually with formol calcium) followed by suitable freezing.

Preparation of Fresh Tissue (Unfixed)

To produce cryostat sections of good quality two criteria must be satisfied. The tissue must be fresh; sections produced from *autopsy material* are *inferior* to those cut from fresh surgical material. Equally important is the initial freezing of the tissue; the rate of freezing is critical.

The faster the *quenching*, the better the tissue will be preserved. Slow freezing of tissue will produce ice crystal artefact in some tissues. The water in the tissue acts as the *embedding* medium, giving the tissue *hardness* when *frozen*. For this reason tissue frozen by one of the *techniques* using very low temperatures should be allowed to warm up to a suitable cutting temperature before *trimming*.

Tissue may be frozen in the following ways.

(*i*) Solid carbon dioxide ('*cardice*')

(*ii*) Cardice mixtures, e.g. cardice/acetone

(*iii*) Carbon dioxide gas (from a pressure cylinder)

(*iv*) Liquefied gases
(*v*) Cold contact
(*vi*) Aerosol sprays.

It is necessary to freeze the tissue as rapidly as possible to avoid *ice crystal* formation. The rate and level of production of ice usually vary with the water content of the tissue to be *frozen*.

Solid Carbon Dioxide (Cardice)

This is one of the *slowest satisfactory* means of *freezing*. The tissue is placed onto a block holder with a drop of water and two pieces of cardice are held against the sides of the metal block holder. Freezing occurs by conduction and a white line can be observed passing through the tissue until it is all frozen.

This method, despite its slowness, is popular in routine histology laboratories when the cryostat is only used for urgent biopsies from the operating theatre. *Cardice* is supplied in 28-lb blocks. The lower the temperature of *storage*, the longer the cardice will last, but with storage at 4°C two blocks will give a constant supply for one week.

Cardice Mixtures

This technique is similar to the previous one, except that the cardice is *crushed* into a fine powder. It is then placed in a *Thermos flask* and an *organic solvent* is added until a slush is obtained. This will remain all day in a *vacuum flask*, allowing many blocks to be frozen.

The freezing technique is to lower the metal block holder into the slush until only a ¼ in is clear and then place a drop of water and the tissue on the *block holder*. Care must be taken with this method, especially if histochemical techniques are going to be applied to the sections, so that the block does not become contaminated with the solvent.

Carbon Dioxide Gas

The freezing of tissue by carbon dioxide gas is carried out by using a piece of equipment like the *Slee bench freezer* which is designed to take the block holders on a small *platform* under which the gas is fed from a cylinder. The block holders must have *holes* on each side for the gas to pass through. The freezer has a *Perspex lid* from which the gas is returned to aid the rapid freezing of the block. It is possible to adapt a conventional freezing microtome for the same purpose, although freezing is slower.

Liquefied Gases

This method is the most rapid of those available to the majority of laboratories *practising* histochemistry. The most readily available gases are *liquid air* or *liquid nitrogen*, and a *Thermos flask* filled with either will usually last for a full day.

Liquid nitrogen is the safer of the two. If tissue needs to be frozen daily it is probably worthwhile to have a supply of liquid nitrogen delivered to the *laboratory*. A suitable container is available to hold 25 litres; taking into account the loss of nitrogen during *storage* and *filling* one Thermos flask a day, this amount would last for almost two weeks. The cost of liquid nitrogen itself is small.

The freezing of tissue in liquid nitrogen is rapid enough for most purposes. The material can be frozen either directly onto the block holder, or by placing the tissue onto a piece of *aluminium foil* and plunging both into the liquid nitrogen. If the second technique is used, the frozen tissue can be stored in *polythene bags* in the deep freeze until required.

A difficulty of freezing tissue with liquefied gases is the cracking of the block. The larger the piece of tissue, the more likely is cracking to occur, although *fibrous* types of tissue tend to crack less than softer ones like liver. With large blocks and soft tissue (liver etc.) a little practice is necessary before this cracking can be avoided.

Cold Contact

The technique suggested by *Cunningham et al.*, involving the *pre-chilling* of glass tubes to –70°C and below, is a useful method. The tubes are kept corked at this temperature and when the tissue is ready for freezing the cork is removed and the tissue placed against the inside of the tube. Freezing is rapid and the method is suitable for almost all *tissues*, *muscle* being the exception.

An advantage of using this technique is that a ready means of storage is available for material which is not required for immediate cutting. Within this group falls the technique utilising metal cooled by the refrigeration coils of the cryostat. In some instances these do not give low enough temperatures to produce rapid and adequate initial freezing of some tissues.

Aerosol Sprays

Recently some commercial firms have marketed aerosol freezing sprays containing gases *liquefied* under pressure e.g. *dichlorodifluoromethane*. Tissue from urgent biopsy specimens can be satisfactorily frozen on the block holder by this technique, although the freezing is neither rapid nor even enough for most histochemical methods. The big advantage of these *sprays* is their *simplicity* and *easy* availability, with minimal storage problems.

Freezing of Muscle

Muscle is the most difficult of routine *biopsy tissues* to freeze without serious artefact, and to obtain satisfactory results three conditions must be fulfilled. First, the piece of muscle selected must be small (not exceeding 5 mm × 4 mm). Secondly it must be left for 10-25 minutes in a moist atmosphere to avoid subsequent contraction *artefact* and, finally, it must be frozen by the most rapid means, usually liquefied gases. Difference when two different freezing techniques are used. Other tissues are satisfactorily frozen by any of the above techniques and if a choice has to be made between them, the final decision must take into account availability of the freezing reagent.

Preparation of Fixed Tissue

For the *demonstration* of hydrolytic enzymes and other substances affected by diffusion, prefixation in 10 per cent formol *calcium* at 4°C is advised. This will ensure more precise *localisation* of the enzyme than by freezing the tissue unfixed. Tissue to be treated in this way is fixed for 18 hours in cold formol calcium, followed by a similar time in gum sucrose. It is important that after fixation and subsequent treatment the tissue is blotted as dry as possible before freezing.

The excess water in the tissue will make the block very hard when it is frozen. Contrary to the need for rapid freezing of unfixed tissue, fixed blocks should be frozen relatively slowly to avoid tissue damage when this excess of water is frozen; hence liquid nitrogen is not recommended. The sectioning of fixed tissue is considerably more difficult than that of unfixed material. This is due to the extra hardness of the tissue block. Block temperatures of below –20°C will cause damage to the microtome knife.

The temperature of the *cryostat* should be raised to between –5°C and –10°C depending upon the tissue; this is usually sufficient for suitable sections to be cut. In cases of difficulty further cooling of the knife with cardice will probably help. Sections cut from tissue treated in this way have a *tendency* to float off the slide or coverslip; to overcome this, the slides or *coverslips* should be treated with a 2 per cent *gelatin-formaldehyde* solution.

Cryostat Temperature

This is of vital importance for the consistent production of thin, *artefact-free* sections. The majority of the standard cryostats have *thermo-*

stats to regulate the chamber temperature between -5°C and -30°C. It is also possible to cool the *knife* and block further with *cardice* to temperatures approaching -70°C.

The use of cardice, however, does not allow for accurate control of the knife or tissue temperature, although arbitrary cooling of the knife with cardice may occasionally enable a satisfactory section to be produced under sub-optimal cryostat conditions. Accurate control of cryostat conditions can only be obtained by using the *thermoelectric* cryostat equipped so that the block and knife temperature can be varied accurately and independently within a few moments.

The development of the thermoelectric cryostat showed that each tissue has its own optimal cutting temperature; kidney and liver will cut best in a conventional cryostat at –15°C to –18°C, while muscle will cut more easily at –23°C to –25°C. *Formalin-fixed* material will cut satisfactorily at –8°C to –12°C, depending upon the tissue. All the above figures are the chamber and knife temperatures. It is important that the tissue should not be too cold during cutting.

If the tissue is below –30°C it will be brittle and the resistance to-cutting such that the block may fracture into many pieces and the knife will be badly damaged. After a block has been frozen it is *advisable* to leave it in the cryostat for 10 minutes to reach chamber temperature, so when urgent *biopsies* are being cut it is better to use carbon dioxide for the initial freezing to minimise this waiting period.

If the unfixed block is too warm the sections will stick to the top of the knife in a wet mass. If the anti-roll plate is too warm, the sections will stick to that also. Most tissues, however, will cut successfully at –15°C to –22°C.

Anti-Roll Plate

The *anti-roll plate* (or *guide-plate*, as it is sometimes called) was originally a piece of glass, usually a *microscope slide*, with two strips of *Sellotape* running down the vertical edges of the plate. The two strips of Sellotape cut from a standard roll are approximately 70 microns thick. The plate is sprayed with *polytetrafluoroethylene* (P.T.F.E.) to reduce the friction between the plate and the section as it is being cut.

The anti-roll plate is carried in a holder which is usually attached to the *microtome*. In the newer models of the *Bright cryostat* the anti-roll plate is a shaped piece of perspex with spacer screws instead of Sellotape. It is possible and necessary to be able to adjust the angle, height and alignment of the plate to the knife.

To produce satisfactory, *crease-free*, flat sections it is vital that the operator of the *cryostat* fully understands the use of the anti-roll plate and is capable of adjusting it correctly. All types of *anti-roll* plate are now micro-adjustable. The tendency of sections of all tissues, when they are being cut in the cryostat, is to curl upwards.

The anti-roll plate must be in a position to guide the section downwards and to leave it resting on the knife face. The two strips of *Sellotape* rest against the knife when the anti-roll plate is in position, the thickness of the Sellotape (or *spacer screws*) allowing a sufficient gap for the section to pass down. For the production of good sections, the following five points should be observed with the anti-roll plate.

(*a*) Anti-roll plate cold (cabinet temperature)

(*b*) Anti-roll plate firmly held in holder

(*c*) Sellotape cut flush with top of anti-roll plate (if Sellotape used)

(*d*) Top edge of anti-roll plate not damaged

(*e*) Good coating of PTFE on anti-roll plate (if glass plate used).

Setting the anti-roll plate in the correct position can cause a new operator considerable trouble, but with practice it is possible to fit a new *anti-roll* plate in the correct position within 30 seconds or so.

No set position can be claimed for the *anti-roll* plate as the angle and height will vary according to the angle at which the knife is set in the *microtome* and the method of *sharpening* the knife. General conditions for setting the anti-roll plate are that the top of the plate appears fractionally above the knife and the angle of the anti-roll plate to the knife is 20°.

This angle applies to a new knife. After adjusting the anti-roll plate, it is important to allow the plate to cool down to the cabinet temperature before attempting to cut sections.

Microtomes

The early British cryostats employed the Cambridge rocking microtome. This microtome was ideal from a *mechanical point* of view at sub-zero temperatures, as no use is made of sliding parts. It also needed only minor alteration to fit the *anti-roll* plate.

The disadvantages of this *microtome*, however, are the limitation of the size of block that can be cut, and the problems of cutting hard pieces of tissue. This led to the choice of rotary microtomes for these *cryostats*. In the U.K. these are rotary rocking microtomes, either the Cambridge or the Slee, both of which incorporate many of the mechanical advantages of the Cambridge rocker, but at the same time are capable of cutting larger and tougher blocks more easily. The *Slee rotary* microtome has a retracting stroke which makes it ideal for serial sectioning of frozen tissue. Of the cryostats produced in America the majority are top-opening, compared with the British front-opening models.

The American cryostats usually employ a more *sophisticated* type of rotary microtome, which from a mechanical point of view is less suited to operating at low temperatures but is more rigid and provides better support for difficult tissues.

The two major British manufacturers both produce cryostats that house the Leitz base sledge; these models are used for *neurohistology*, *neurohistochemistry*, bone sections and whole body (rat) *autoradiography*. Due to the weight of the microtome, they need to be motor-driven. This allows for an infinitely variable cutting speed which enables many types of tissue to be cut.

The application of motor-drive to the cryostat is also applicable to the rotary microtomes and this type of cryostat is a decided advantage when cutting serial sections from a block; this refinement is probably not necessary in a routine surgical laboratory.

Microtoine Knives

Any knife designed for the microtome in the cryostat will cut satisfactory sections when new. The difficulties occur when the knife needs *resharpening*. The fact that the anti-roll plate has to be adjusted against the knife edge raises problems in regard to sharpening the knife. If the knife is frequently *honed* by hand, the facet on the knife does not allow the anti-roll plate to be correctly adjusted.

This difficulty can be delayed by using a knife back, but the problem remains. The use of *automatic* knife sharpeners is better, but these also produce a facet. In the author's experience, the best results are obtained by having the knives reground. The knife, when in need of sharpening, is removed from the *cryostat*, dried and sent away to be reground.

The cost is not high and when the knife is returned it is ready to be placed into the *cryostat* without any further *attention*. If the knife is stropped it has a fine edge, which is soon removed by the hard frozen blocks. Unless very thin sections are required it is advisable not to treat the knife in any way after *regrinding*.

Method 5.1

Preparation of tissue for prefixed cryostat sectioning

(*a*) Remove tissue as rapidly as possible from the body

(*b*) Cut into pieces 1 cm × 1 cm × 0.5 cm
(*c*) Place into 10 per cent formol calcium at 4°C overnight.
(*d*) Wash in running tap water for 5 min
(*e*) Place in gum sucrose solution for 24 h
(*f*) Blot as dry as possible
(*g*) Freeze tissue and mount on block holder.

Method 5.2

Rapid staining of cryostat sections (haematoxylin-eosin)

(*a*) Fix sections in formol-acetic-alcohol for 40 s
(*b*) Rinse rapidly in tap water
(*c*) Stain in Harris haematoxylin at 37°C for 30 s
(*d*) Rinse rapidly in tap water
(*e*) Differentiate in 1 per cent acid alcohol
(*f*) Rinse in Scott's tap water
(*g*) Rinse rapidly in tap water
(*h*) Stain in 1 per cent eosin for 10 s
(*i*) Rinse rapidly in tap water
(*j*) Dehydrate through graded alcohols to xylene and mount.

Method 5.3

Preparation of lung tissue for satisfactory cryostat sectioning

(*a*) With animal tissue inject 4 per cent gelatin into the lung
(*b*) Remove lungs from animal and place in running cold water (to congeal gelatin) for 5-10 min
(*c*) Slice lung to required size
(*d*) Freeze lung by one of the techniques in other chapter of this book.

If only a small piece of lung is available, i.e. biopsy specimen, infiltrate the lung with as much gelatin as possible before freezing.

Note. Tyler and Pearse recorded no difference between injected and uninjected lungs with the enzyme methods. Using uninjected lung however, only small fragments of section could be obtained.

6

Freeze Drying and Freeze Substitution

In simple terms *freeze drying* is the rapid freezing of tissue, followed by the removal of water under vacuum whilst the tissue remains *frozen*. Freeze substitution is a variant of the freeze drying technique in which the water in the *frozen tissue* is replaced by an organic solvent at sub zero temperatures. Freeze substitution is not carried out under *vacuum*.

The technique of freeze drying is probably the most misunderstood of the *histochemical* processing methods. It is a common error for freeze drying to be regarded as a method of fixation; this is not so. The fixation of freeze dried material can be accomplished by two means :

(*a*) vapour fixation of the block before *impregnation* with wax; and

(*b*) fixation of the embedded sections before *staining*. A comparison is sometimes made between freeze dried sections and *unfixed* frozen (*cryostat*) sections.

The important difference in freeze dried material, apart from the embedding in paraffin wax, is that the disruptive effects of thawing of the frozen sections is avoided, all the unbound water having been removed while the tissue is below freezing point. The importance of this fact is frequently overlooked in the appraisal of these two techniques.

REASONS FOR FREEZE DRYING

The standard methods of tissue preparation involving fixation, *dehydration*, clearing and embedding in *paraffin wax* have a number of drawbacks for the *histochemist*. These procedures affect tissue by causing loss of some soluble substances, by the alteration of chemically reactive groups and by the *displacement* of cell constituents caused by the penetration of the various fluids.

Some of the advantages obtained with freeze drying are, however, lost when the freeze dried tissue is embedded. This applies particularly to enzymes, especially the *oxidative* types. For the critical *demonstration* of *carbohydrates*, *proteins* and some *hydrolytic enzymes* freeze drying is ideal.

FREEZE DRYING TECHNIQUE

This involves the tissue being frozen rapidly to preserve its *chemical composition* and cell structure much as it was in life. The *unbound water* is removed from the tissue by means that cause little or no change to its chemical composition. Once the tissue is dry it is embedded and sectioned. The technique of freeze drying can be divided into four parts.

(1) Freezing the tissue (*quenching*)
(2) Drying the tissue (*dehydration*)
(3) Fixation
(4) Embedding and sectioning.

Quenching

It is necessary to quench the tissue rapidly, for the more rapid the quenching, the smaller the *ice-crystals* formed and, the smaller the ice-crystals, the more the tissue will retain its original appearance. Many workers have tried to avoid this ice-crystal artefact by soaking the tissue in different reagents before freezing, glycerin and gelatin being the most popular, but without much success.

To obtain a fast rate of cooling, a *quenching liquid* with a high rate of thermal conductivity is required. This fact has been stressed by many workers, including Simpson. The requirements are for a solution of high thermal conductivity and with as low a temperature as possible. If a solution with low thermal conductivity is used, a layer of *vaporised gas* is formed around the block, slowing down the transference of heat from the tissue to the freezing medium.

In recent years many different solutions have been tried, *Gersh* used liquid air at -195°C. Although this will freeze the tissue at this *temperature*, it has, unfortunately, a low thermal conductivity. Other agents tried include ethyl alcohol cooled by liquid air, *isopentane* cooled by *liquid nitrogen* and pentane cooled to -125°C. Bell recommended the use of *dichlorodifluoromethane* (Arcton 12), the coolant used in domestic refrigerators. The gas is condensed into a beaker, and the beaker containing the liquid Arcton is then placed carefully into a flask of liquid nitrogen. When the Arcton freezes the beaker is removed from the flask, and as the Arcton starts to thaw (at -158°C) the tissues for freeze drying are immersed in the liquid. A recent advance is to use *Arcton 22* (*Freon 22*), recommended by Rebhun and Pearse as the fastest quenching agent. It is used in the manner described for *Arcton 12*. Other workers prefer different liquids. *Moberger*, *Lindstrom* and *Anderson* used pure propane cooled with liquid nitrogen.

One problem with quenching is that certain tissues will crack, due to differential expansion of the tissue during freezing. Liver is especially vulnerable and care must be taken when quenching large pieces. It is advisable to quench the tissues by placing them on a strip of *aluminium foil* and *immersing* them in the liquid.

Eranko pointed out that a much faster rate of cooling could be achieved by placing the tissue between two strips of copper previously cooled by *immersion* in liquid nitrogen. However, the majority of blocks for freeze drying are *irregularly* shaped so this method cannot often be used.

Dehydration or Drying

This phase of the freeze drying technique consumes virtually all the time involved. If a slice of fresh *kidney* weighs 100 mg there will be 75-80 mg of water which must be removed during the drying phase. A *satisfactory* means of removing this large amount of water has to be employed without damage to the tissue.

Drying of the tissue will occur when heat is supplied to the ice crystals in it, thus causing water molecules to vaporise from their free surfaces. According to *Stephenson*, after the molecules *sublime* they pass either into the *vacuum chamber* through the dry shell of the tissue, or they return to the interface between the dry shell and the frozen wet tissue. Although both processes may take place *simultaneously*, it is considered that the majority of the water molecules pass into the vacuum chamber to be absorbed by the vapour traps.

The drying phase is carried out in a vacuum at a higher temperature than that of the initial quenching. If the temperature of the tissue is maintained at -40°C in a vacuum of 10.3 torr, or better, then under these conditions, relative heat is being supplied to the tissue and the pressure of the environment around the tissue is lower than the vapour pressure of the ice, so that the latter sublimes, while the tissue remains solid and the growth of *ice crystals* is delayed. According to Pearse, the temperature of the tissue during the drying phase is critical. He suggested that -30°C to -40°C is the best temperature range since ice *vaporises* ten times more slowly at -60°C than at -40°C. If a reasonably short *drying* time is required, it is important not to have the tissue too cold, although care must be taken to ensure that the *temperature* during drying does not rise above -30°C, When the ice sublimes from the tissue it leaves a dry outer shell.

Through this dry shell of tissue the remaining ice has to pass during *sublimation* and must not damage the extremely delicate tissue of the already dried outer shell. The relative heat necessary to complete the drying of the tissue must pass through this *dry shell*. Unfortunately, the dry outer shell is a poor conductor of heat but nevertheless *relative heat* must be continually supplied to the wet core of the tissue.

The amount of heat supplied must be carefully controlled since too much heat will cause damage to the dry shell of the tissue and too little heat will not allow efficient drying of the remaining wet core. When the water molecules have left the dry shell of the tissue and have entered the vacuum chamber, they are removed by a vapour trap. These traps are usually one or more of two main types.

(*a*) A chemical trap (e.g. *phosphorus pentoxide*)

(*b*) A cold trap (which may be a liquid gas, e.g. nitrogen or air *or* a mixture of *solid carbon dioxide* with *alcohol* or *acetone*).

A common *chemical trap* employs phosphorous pentoxide which is a desiccant and when placed into the vacuum chamber will absorb water molecules reaching it. The container is usually placed between the tissue and the vacuum pump line.

The second type of trap is usually called a '*cold finger*'. It consists of a piece of *Quickfit glassware* shaped like a finger. The '*nail*' end of the finger is usually placed about 5 cm from the tissue. The cold finger is filled with liquid nitrogen or other *refrigerating* fluid and placed into the drying chamber some 5 cm away from the tissue. The water *molecules* passing from the tissue attach themselves to the cold finger.

There are many opinions about the correct distance of cold fingers from the tissue during the drying phase. Most of this literature is complicated, involving unconfirmed complex mathematical theory, which is beyond the scope of this book.

Many attempts have been made to reduce the drying time of the pieces of tissue. These attempts have been *frustrated* mainly for two reasons. First, the drier the tissue becomes the more difficult it is to remove the remaining water molecules through the dry shell, which provides a strong resistance to them. Secondly the amount of heat that can be supplied to the piece of tissue during the drying phase is strictly limited, because the temperature of the tissue must not be raised above -30°C. If it is, damage will occur to the dry shell of the tissue.

Workers in the past have carefully dried blocks for 12-48 hours maintaining the heat to the tissue at -40°C to -50°C, and then for the last hour they have raised the temperature from -40°C to -10°C. It is in this hour that damage due to *ice crystal* formation may occur, because there is still a considerable amount of water remaining. It is *imperative*, then, that the drying of the tissue be carried out at the same temperature throughout, preferably between -35°C and -55°C.

The temperature should not be allowed to rise above -30°C with the vacuum pump operating unless the *operator* is in no doubt that the tissue is completely dry. When the tissue block is dry, it is possible to remove the block from the *freeze dryer* without causing any apparent damage to it. If the tissue is completely dry it can be left at room temperature and in the normal *laboratory* atmosphere without any outward change.

A piece of tissue that is only partially dry, however, will start to rehydrate as soon as it is removed from the freeze dryer and left in the laboratory atmosphere; this water probably spreads outwards from the undried core.

Fixation

For the majority of histochemical techniques freeze dried tissue requires subsequent fixation. The method of choice is to fix by *vapour fixation* of the freeze dried block, although sections may br fixed after suitable embedding and cutting. *Fixative vapours* may be introduced into the drying chamber of the freeze dryer, but this is not recommended.

The best results are obtained by vapour-fixing the block in a special piece of *equipment* in which it is possible to vary the temperature and moisture content of the surrounding atmosphere. Control of the moisture content is particularly important for the demonstration of catecholamines in the tissue.

A number of substances are suitable for use as vapour fixatives, notably formaldehyde, glutaraldehyde and osmium tetroxide. The easiest to use, and the most successful, is *formaldehyde* vapour which gives excellent preservation of the tissue and good *localisation* of most substances, particularly carbohydrates and proteins. Formaldehyde vapour is produced by heating paraformaldehyde powder at 60°C .

An alternative method of fixation is to fix the sections cut from the unfixed embedded block, using any of the standard *aqueous fixatives*.

The accuracy of localisation of the substances to be *demonstrated* is usually much inferior to that obtained with vapour fixation.

Embedding

When the tissue has been dried, and possibly vapour-fixed, it is ready for the embedding stage of the process. Many different *embedding* media can be used, the most popular being paraffin wax, but the choice depends on the structures and substances to be demonstrated. *Lipids* and *enzymes* are usually better demonstrated after embedding in polyester wax or even *Carbowax* despite some problems associated with infiltration of *Carbowax* into dry tissue.

In many freeze dryers the embedding process takes place within the freeze dryer without the vacuum being broken. The *paraffin* wax must be placed in the dryer before the tissue and degassed under vacuum. When freeze drying is complete, the temperature is raised from -40°C to one or two degrees above the *melting point* of the paraffin wax. The tissue is then impregnated with the wax whilst still under vacuum. This technique, while *admirable* in many respects, has the disadvantage, in the author's experience, of occasionally causing damage to the tissue.

As long as the tissue is dry, the best policy is to remove it from the dryer after it has reached room temperature and then either place it immediately in degassed paraffin wax in a normal histological vacuum *embedding* bath or expose to vapour fixation, followed by paraffin wax embedding. The tissue is then embedded in fresh wax and is ready for cutting.

Cutting of Paraffin-Embedded Freeze Dried Blocks

This presents only minor difficulties to an experienced histology *technician*. The blocks, fixed or unfixed, tend to be more brittle than routine processed paraffin blocks, and should be embedded in paraffin wax with a high content of *plastic polymer*. Cooling of both block and knife is usually sufficient to overcome the problem caused by the brittle nature of the block.

From properly impregnated paraffin blocks of freeze dried material- it is possible to cut serial sections of 3 microns with little difficulty. It appears more difficult to cut unusually thick sections than thin ones, due probably to the brittleness of the block.

Floating out of Sections of Unfixed Freeze Dried Blocks

This presents much more of a problem than the cutting of the sections. Because of their *hygroscopic nature* unfixed sections cannot be floated out on to a *waterbath* in the normal way. Sections cut from blocks of *vapour-fixed material* can be left on warm water for a limited time before being picked up on slides. The most common floating out solutions for unfixed sections are

(*a*) 70 per cent alcohol at 40-50°C
(*b*) 10 per cent formol saline at 45°C
(*c*) 10 per cent formol calcium at 45°C
(*d*) 18 per cent sodium sulphate at 45°C
(*e*) warm mercury at 40°C (toxic)
(*f*) dry mounting on a warm slide.

Each of these techniques has its own disadvantages. The first three allow perfectly flat sections to be picked up on slides, but the sections are subject to fixation and, in the case of 70 per cent alcohol, to loss of some *lipid substances*. The floating out of sections on warm mercury (*warm mercury* is *toxic* and care must be taken) involves no fixation or loss of any substances; the method is more difficult than most *floating* out *procedures* and some practice is required before *repeatable* success is obtained.

An important factor is to keep the mercury as clean as possible. This means frequent *filtering* if many sections or blocks are to be cut. The method of floating out on mercury was first used by Harris, Sloane and King. The last method, favoured by many workers, is to coat microscope slides with *glycerin albumen* and to place them on a hotplate at 45°C. The section is cut, placed on the dry warm slide, and then flattened by finger pressure. Whilst this method causes least damage to the section, crease-free sections are rarely obtained.

Furthermore, it is more difficult to keep the sections from floating off the slide during the subsequent staining technique. Careful consideration must be given when deciding which *floating* out method to use, and it may be necessary to use more than one method while cutting the same block, in order to obtain the best possible results from the stains or reactions to be carried out.

Embedding in Media other than Paraffin Wax

Embedding in water-soluble waxes, the most common of which is *Carbowax*, has been used with *freeze dried blocks*. The main advantage is that no heat and no *organic solvents* are required for removal of the wax. The wax can be removed from the slide with water. The use of this method in routine *histology laboratories* has been restricted because of the difficulties of floating out the sections. Water cannot be used, and removing the creases is also a problem. With freeze dried blocks the same difficulties occur and, in the author's experience, difficulty is also encountered in getting the Carbowax to penetrate into completely dry tissues. If the block of tissue is slightly wet the Carbowax will *penetrate*.

As an alternative to embedding in paraffin wax, it is possible to double-embed blocks. This technique, takes considerably longer, but if the substance to be demonstrated will resist the procedures involved, it is possible to obtain sections of a higher standard than can be obtained by any other method. The method is given in full at the end of this chapter . It involves processing through *alcohols* and *celloidin* at deep-freeze temperatures before finally embedding in paraffin wax. An advantage is that sections can be floated out onto warm water in the usual manner. Burstone used this method to demonstrate hydrolytic enzymes in freeze dried material.

An *acrylic resin* such as methacrylate, used much as the electron microscopists use it, is sometimes preferred as an embedding medium when thin or very thin sections are required. *Polymerisation* of methacrylate or other resins is achieved by heating at 50°C for eighteen hours or at 60°C overnight, and for this reason, the method is only used for specialised techniques.

Applications of Freeze Dried Material

Freeze dried material has many applications. Pearse placed them under the following four main headings.

(1) Examination of unfixed material in inert media by *physical* methods

(2) Examination of unfixed material by *microincineration* or by treatment with buffer extractions, etc.

(3) Examination of suitably fixed or unfixed material by histo chemical and conventional histological methods

(4) Examination by *autoradiographic* techniques.

Unfixed material in inert media

Unfixed freeze dried sections can be examined by phase contrast microscopy, polarised light, dark-field illumination and fluorescent microscopy.

Extraction with buffer solutions and microincineration

The proteins of freeze dried sections are said to be almost completely undenatured and therefore it is possible to test their solubility in different buffer solutions. Following the extraction, the section is fixed and a suitable reaction for the protein may be carried out alongside a control section to help establish the quantity extracted.

Microincineration is a technique carried out to determine the presence of inorganic elements in the tissue sections. It is important that fixation and processing of the tissue must not alter its mineral content. Freeze dried sections are therefore ideal for this purpose.

Fixed and unfixed sections and staining techniques

The choice of a fixative with freeze dried sections is dictated by the components to be demonstrated or the particular histological staining method to be applied. All the routine histological stains can be used on freeze dried sections, although the staining times must be reduced, possibly by one third for the majority of methods.

It is also possible to demonstrate enzymes, but the use of the heat, paraffin wax and a combination of the two, reduce the degree of enzyme activity and its localisation is open to question. For *mucin* and other *mucopolysaccharides* this is probably the best means of demonstration. A comprehensive study was carried out by Tock and Pearse and readers are referred to this for details.

Examination by autoradiographic techniques

Freeze dried sections are worth consideration when water soluble *isotopes* are to be used and their usefulness should be compared with that of frozen sections.

HISTORY AND TYPES OF FREEZE DRYERS

There have been many different types of freeze dryers developed over the years. The drying times of the tissues have varied from a few hours to many days and the slices of tissue dried in some cases were very small indeed. Mann described that Leewenhoek, one of the founders of microscopical science, prepared samples of muscle by a method of freeze drying.

However, the method was not really exploited until the end of the 19th century when liquefied gases became available. A method which involved drying small fragments of tissue *in vacuo* over concentrated sulphuric acid. After several days it was possible to remove the tissue and embed it in *paraffin wax*. Altman stated that this method eliminated histological *shrinkage*.

The second major advance in freeze drying technique was made by Gersh. His equipment, with its various modifications by many people, including Stowell, Jensen and Burstone, is basically the equipment that is still used in laboratories today. Gersh used a Dewar flask containing *liquid ammonia*, which was *hermetically sealed* to the drying chamber.

The temperature inside the chamber was -22°C. To obtain the vacuum, two pumps were used, a rotary pump and a mercury vapour diffusion pump. He used a *phophorus pentoxide* vapour trap. He also incorporated a compartment for the *vacuum-embedding* of the tissue.

Two years later Goodspeed and Uber, instead of using liquid gases as a means of cooling the tissue, placed the drying tube in a mechanically refrigerated chamber at -29°C. A single mechanical pump was employed and *phosphorus pentoxide* was used in the vapour trap. *Goodspeed* and *Uber removed* the tissue from their dryer before *wax-embedding*.

A method of freeze drying based on the use of a dry gas moving over the frozen tissues to remove the water *evaporating* from them was reported by Treffenburg, Jansen and Kramer and Hill. Some provision for drying the air and maintaining the low temperature of the air and specimens was required during dehydration. Kramer and Hill used a mixture of carbon dioxide snow and *cellulose* in a vacuum flask to cool the chamber and to supply a stream of dry gas. The drying time claimed for this type of freeze dryer is 10-24 h for small pieces of tissue.

The next significant advance in the field of freeze drying was made by Pearse who designed a freeze dryer to operate with thermoelectric modules. This freeze dryer was subsequently produced commercially by Edwards High Vacuum Ltd. The tissue is kept at the required temperature throughout the drying phase by the use of *thermomodules*. At this point, before describing *thermoelectric* freeze drying, a brief explanation of how a *thermomodule* works may be necessary.

HOW A THERMOMODULE WORKS

The thermomodules produce cold by means of a phenomenon known as the *Peltier effect*. When a direct current is passed across the junction of two dissimilar metals, heat is emitted or absorbed according to the direction of the current. Recent advances in the technology of *semiconducting* materials have led to the production of *thermoelements* with considerable *refrigerating* capacity.

The modules consist of an array of *bismuth telluride* couples, connected in series by copper strips which serve both as electrical conductors and for heat transmission. When a current of direct electricity is passed through the module, heat is absorbed on one side, and evolved on the other. *Anodised* aluminium plates are bonded to the faces of the module.

With the hot face of the module at 27°C, the cold face of the module is capable of attaining -30°C. The cooling produced by the module is dependent upon the amount of the operating current, thereby providing a simple technique for precise temperature control. If the polarity of the electricity supply is reversed, a heating effect can be produced instead of a cooling one.

THERMOELECTRIC FREEZE DRYING

To obtain a satisfactory working range of temperatures, +60°C or more (to melt the embedding wax, if required) to below -60°C, it is necessary to use more than one module. These *thermomodules* are arranged in a *cascade* effect, the base set of modules having more than the top in the ratio of 4:1. This is to allow the heat produced by the warm face of the top set of modules to be satisfactorily removed by the cold face of the bottom set, allowing the top surface of the top module to drop below -60°C if necessary.

The heat from the base set of modules is removed by cold running water. This use of *thermomodules* obviates the necessity for employing liquefied gases or deep freeze cabinets during the drying phase. It also allows more accurate control of the drying temperature by the simple means of controlling the current supply to the modules. The *vacuum pump* and electrical system are set into the base of the cabinet, while the modules are mounted on the top and are covered by a *desiccator lid*.

The area covered by this *desiccator lid* is the drying chamber. The controls are mounted on the front panel of the cabinet with two gauge dials giving the *module* temperature and the vapour pressure. The controls allow the operator to isolate the drying chamber from the vacuum pump and to regulate the temperature of the *thermomodules* from + 60°C to -60°C.

The vapour trap usually used in the freeze dryer is *phosphorus pentoxide*, contained in two boats placed near the vacuum line. Although the temperature of the module is held at -40°C during freeze drying, there is a temperature gradient above the tissue and at the *desiccator* lid it reaches approximately +15°C. The temperature difference between -40°C to +15°C facilitates the drying of the tissue to such an extent that in four hours a 4 mm slice of tissue at -40°C will be dry.

The majority of other types of freeze dryers need 12 hours or more to complete drying. The block may simply be placed on the module, or it may be firmly attached by ice. Drying occurs faster when the tissue rests on the surface of the module.

In summary, freeze driers may be of the following types.

(*a*) Vacuum chambers cooled by liquid or solid refrigerators

(*b*) Vacuum chambers cooled by mechanical refrigerators

Table 6.1: Drying times using thermoelectric freese dryer.

Thickness of tissue (mm)	*Drying temperature (°C)*	*Drying times (h)*
1	- 30	1½—2
1	- 40	2—2½
1	- 50	3—4
1	- 60	6—6½
2	- 30	3—3½
2	- 40	4—4½
2	- 50	5—6
2	- 60	8—10
3	- 30	4—5
3	- 40	5—5½
3	- 50	7—8
3	- 60	Dry overnight
4	- 30	5—6
4	- 40	6½—7
4	- 50	Dry overnight
4	- 60	Dry overnight

(*c*) Refrigerated drying tube through which dry gas passes

(*d*) Thermoelectric.

The first three types are now largely of historical interest in Britain. The thermoelectric type with its accurate temperature control and its efficient drying ability is now the most widely used type of freeze dryer. For a detailed comparison of the *theoretical aspects* of the different freeze dryers readers are referred to Pearse.

FREEZE SUBSTITUTION

The technique of freeze substitution was originally described by Simpson as an alternative technique to freeze drying. The tissue is frozen as rapidly as possible, and the ice replaced or '*substituted*' by leaving the tissue in *dehydrating* solutions (usually an *organic solvent*) at sub zero temperatures. The technique can be divided into three stages.

(*a*) Quenching or freezing

(*b*) Substitution

(*c*) Embedding.

Quenching

The tissue for freeze substitution should be absolútely fresh and frozen as rapidly as possible. To obtain good results, the quenching must be as fast as possible, for, as with freeze drying, damage by ice crystal formation will ruin the end result. The blocks to be substituted should be small and frozen in a bath super-cooled to 190° by liquid nitrogen. The technique described in Method 6.1 for freeze drying is ideal for this technique.

Substitution

The ice in the tissue can be substituted in two types of solutions either (a) a pure dehydrating solution or (b) a *dehydrating* solution containing a fixative. Consideration must be given in the latter case to the effect of the fixative at the substituting temperature. The use of a fixative such as *picric acid*, *mercuric chloride* or 1 per cent *osmium tetroxide* in the dehydrating solution is normally used for morphological studies and not for critical histochemical work. Substitution has been carried out in a wide range of temperatures from -20°C down to -130°C.

The majority of the work has been carried out at around -70°. The temperature at which the *substitution* is performed is important for a number of reasons; the lower the temperature the less the growth of ice crystals, and the lower the temperature the slower the *penetration* of the substituting fluid. A number of solutions may be used to substitute the ice in the tissue, the two most common being ethanol or acetone.

The penetration of the dehydrating solution will normally take up to about six days at -70°C. *Agitation* during substitution should be employed as often as possible.

Embedding

When the tissue is thought to be fully dehydrated it is transferred to a refrigerator at 4°C. When the substituting fluid has reached this temperature the tissue is immersed in chloroform at 4°C overnight and then in chloroform at room temperature before embedding in either paraffin wax or polyester wax.

Section freeze-substitution

As more sophisticated techniques are developed with the *cryostat*, section freeze-substitution has become a simpler and less *time consuming* procedure than block substitution. The tissue is frozen as for freeze drying, sectioned at as low a temperature as possible, and the sections are transferred direct to cold acetone at -70°C overnight. From the acetone they are picked up on slides and allowed to dry. For the demonstration of oxidative enzymes the sections should be coated with *celloidin*.

The advantage of freeze-substituted cryostat sections over normal cryostat sections is that the water is removed without the sections thawing and thus avoiding the diffusion which usually occurs during *thawing*; it must be remembered that diffusion will occur during the incubating phase of the *histochemical* technique. Whether the results obtained by section freeze-substitution are superior to the standard prefixed cryostat sections is a matter for debate.

Method 6.1

Quenching-equipment required

Dewar flask full of liquid nitrogen or liquid air
Cylinder of Arcton 22 or Freon 22 (I.C.I. Ltd.)
50 ml Pyrex beaker
Copper discs or aluminium foil Pair forceps
Scalpel

(*a*) Obtain fresh tissue and slice into pieces no thicker than 5 mm
(*b*) Into a 50 ml Pyrex beaker run 20 ml Arcton 22
(*c*) Place beaker into a Dewar flask containing liquid nitrogen
(*d*) Take care not to let the beaker slip below the surface of the nitrogen
(*e*) Wait for the Arcton 22 to freeze (5 min or less)
(*f*) Remove beaker of frozen Arcton 22 and stand on bench
(*g*) Place tissue on copper disc or strip of aluminium
(*h*) As Arcton 22 begins to thaw plunge the tissue on the strip or disc into Arcton for 40 s
(*i*) Remove tissue from the beaker of Arcton 22 and transfer to the freeze dryer

Method 6.2

Quenching

(*a*) Liquid nitrogen in Dewar flask
(*b*) Place two copper discs into flask for 2 min
(*c*) Remove discs from coolant
(*d*) Place tissue between the discs for 30 s
(*e*) Remove tissue and place in freeze dryer

Method 6.3

Operating the thermoelectric freeze dryer (with waxembedding outside freeze dryer)

(*a*) Switch on vacuum pump
(*b*) Turn on tap water
(*c*) Switch on current to thermomodules set to maximum
(*d*) Fill containers with phosphorus pentoxide
(*e*) Check that temperature is below -40°C
(*f*) Break vacuum by isolating pump and removing desiccator lid
(*g*) Place containers of phosphorus pentoxide in chamber
(*h*) Place tissue previously quenched by Method 8 or 9 on the module surface
(*i*) Evacuate chamber by opening isolation valve
(*j*) Set module temperature (usually —40°C to -60°C)
(*k*) Check (*i*) temperature is correct; (*ii*) vacuum reading correct
(*l*) Leave to dry (see Table elsewhere in this chapter for approximate time)
(*m*) When dry, isolate chamber and switch off vacuum pump
(*n*) Switch off current to modules
(*o*) Turn off running water
(*p*) Let tissue reach room temperature
(*q*) Remove tissue from drying chamber
(*r*) Place in molten degassed paraffin wax or
(*s*) Fix in vapour as in Method 6.5
(*t*) Embed

Method 6.4

Alternative method of freeze drying to Method 6.3

(*a*) Switch on vacuum pump
(*b*) Set modules to heat and switch current on
(*c*) Place molten paraffin wax in brass boat, and place on module surface
(*d*) Evacuate chamber
(*e*) Check temperature is two or three degrees above melting point of the wax
(*f*) After degassing the wax, reverse current and cool the modules till the wax is solid

(*g*) isolate chamber and remove boat from module surface

(*h*) Remove wax from centre of boat, to enable tissue to be in contact with metal surface and put in tissue

(*i*) Replace boat in freeze dryer

(*j*) Evacuate chamber

(*k*) Continue from Step (*c*) Method 6.3 until Step (*m*)

(*l*) When tissue is dry reverse current and start to heat the boat

(*m*) Watch for paraffin wax to melt then control temperature to 3°C above the melting point of wax while infiltrating the tissue

(*n*) Wait for 10 min

(*o*) Isolate pump

(*p*) Remove tissue and embed in fresh degassed paraffin wax

Method 6.5

Formol vapour fixation of freeze dried blocks

(*a*) Quench as in Method 6.1 or 6.2

(*b*) Freeze dry as in Method 6.3 until Step (*q*)

(*c*) Place tissue in airtight vessel containing a quantity of para formaldehyde and with variable temperature and humidity control between +20°C and +70°C

(*d*) Leave for 1-3 h usually at +60°C

(*e*) Remove tissue and switch off heat

(*f*) Place tissue in molten paraffin wax in embedding bath, for 30 min or longer depending upon size of tissue

(*g*) Embed in fresh wax

Method 6.6

Double embedding of freeze dried blocks

(*a*) Quench as in Method 6.1 or 6.2

(*b*) Freeze dry as in Method 6.3 till Step (*m*)

(*c*) Place tissue in alcohol/ether 1 :1 at -70°C for 3 h

(*d*) Place tissue in 2 per cent celloidin (in alcohol/ether, 1:1) at -70°C for 16 h

(*e*) Excess celloidin removed by blotting

(*f*) Place tissue in chloroform at -70°C for 1 h

(*g*) Place tissue in fresh chloroform at -70°C for 1 h

(*h*) Place tissue in chloroform and allow to reach room temperature

(*i*) Place tissue in paraffin wax for 1-2 h
(*j*) Embed

Method 6.7

Block freeze-substitution

Reagents required

(*a*) Liquid nitrogen or liquid air
(*b*) Arcton 22 or Freon 22
(*c*) Acetone
(*d*) Chloroform
(*e*) Paraffin wax

Technique

(*a*) Obtain small fresh slices of tissue
(*b*) Quench by placing in Arcton 22 precooled by liquid nitrogen
(*c*) Place frozen tissue in precooled acetone -70°C for 2-5 days
(*d*) Transfer tissue to chloroform at -70°C for 2-3 h
(*e*) Allow chloroform to reach room temperature slowly
(*f*) Fresh chloroform at room temperature for 1 h
(*g*) Transfer sections to paraffin wax
(*h*) Fresh paraffin-then embed

Method 6.8

Section freeze-substitution

Equipment required

(*a*) Cryostat
(*b*) Liquid nitrogen or liquid air
(*c*) Arcton 22 or Freon 22
(*d*) 0.5 per cent Celloidin, solution in alcohol/ether 1 : 1
(*e*) Acetone

Technique

(*a*) Obtain fresh slices of tissue
(*b*) Quench by placing in Arcton precooled with liquid nitrogen
(*c*) Transfer block to cryostat and cut sections at as low a temperature as possible
(*d*) Drop the sections in precooled acetone
(*e*) Transfer acetone to -70°C for 18-24 h
(*f*) Mount sections on coverglasses (coat with Celloidin if necessary).

7

Living Chromosomes

The observation of chromosomes in the living state is useful as an introduction to their study, and it also serves certain special purposes of physical research. The optical methods which were formerly of interest for showing that living chromosomes tallied with the appearance of fixed ones are now chiefly used for the study of fixed material.

The most suitable material for direct observation of cell division and chromosome movement is as follows.

Plant

Staminal hairs and young petals, *Tradescantia,* diploid or tetraploid species, Belar, Barbar, Kuwada and Nakamura.

Pollen tubes, many plants, Wulff.

Antheridial filaments, *Chara* and *Nitella.* Karling and rhizoids, *cf.* Decker.

Endosperm, *Iris, Leucojum Haemanthus,* etc. Bajer and Bajer, 1956; Ostergren and Bajer.

Animal

Salivary glands, Diptera, *Chironomus.* Bauer; Poulson and Metz. Testes, Orthoptera (Meiosis), Belar, Nicklas, Nicklas and Staehly.

Tails of Urodela, Barber and Callan.

Methods

Removal of excess endosperm juices in the Bajers' method allows mild flattening of the cells by surface tension, the flattened cells entering mitosis in a normal way.

The mounting media for these tissues are in order of importance

(*a*) Liquid paraffin.

(*b*) Natural juices squeezed out of adjoining tissue.

(*c*) Ringer or isotonic salt solution.

(*d*) Sucrose solutions (especially for pollen tubes, q.v.).

(*e*) Agar with sucrose.

Liquid paraffin may be used to ring the cover-slip and prevent evaporation even where it is not used for mounting. Its value depends on its capacity for dissolving five times as much oxygen as water will. In a class by itself is the examination of salivary glands in whole living larvae of *Drosophila* or *Sciara* in air or young *Chironomus* in water, ringed with vaseline to prevent squashing, and lightly pressed to show the gene bands. To exclude heat radiation from the source of light the beam should be passed through a water bath containing a weak aqueous solution of ferrous ammonium sulphate.

Vital Staining

The alleged vital staining of chromosomes by methylene blue is perhaps staining but *scarcely vital*, since this oxidising agent certainly injures the chromosomes as soon as it becomes effective. The same applies to a number of other vital stains that are said to have been effective.

Chromosome Optics

The chromosomes have three optical properties that make their observation possible without staining.

(1) For the first method of observation it is necessary to cut down the lighting so as to make use of the *differential refractivity* of the chromosomes and the spindle. The simplest method, used by Kuwada and Nakamura. Wada, is to restrict the cone of light passing through the condenser.

Films of chromosome movement can be taken in this way. For the study of the action of fixatives dark ground illumination has been found useful. The principle of this method is to illuminate the object from the edge of the normal cone so that the light entering the objective is chiefly that transmitted by differential refraction. It is not therefore reliable as a means of studying fine structure.

(2) The second method depends on their *differential absorption* of *ultraviolet* light. With an ultraviolet microscope, photographs of living chromosomes have accordingly been taken. The maximum absorption

was found by Caspersson to be at 2600 Å. The greatest value of this method lies in the higher resolving power of light with the shorter wavelength which then can be applied to fixed chromosomes, stained or unstained.

(3) The third method is that of *birefringence* shown by polarised light. It depends on the regular molecular orientation of the chromosomes especially when stretched naturally as in the filamentous grasshopper sperms, or artificially as in accidents of smearing. This property, which they share with the mitotic spindle makes it possible for them to be seen between crossed Nicol's prisms by polarised light.

For this purpose, although living observation is possible, fixation in alcohol has been used since it is naturally more convenient for photographic purposes, and desiccation perphaps intensifies molecular orientation.

All these three properties, like those of differential staining, are presumably due, to the nucleic acid attached to the protein chain of the chromosome.

(4) The fourth method is by *phase-retardation* of light according to Zernike's phase-contrast method. The insertion of phase-changing plates into the optical system of the *microscope* give visible differences in the intensity of light absorption between cell constituents having different refractive indices. Under these conditions the chromosomes from late prophase onwards appear opaque, indeed as opaque as if deeply stained.

(5) The fifth method is by Nomarski differential-inter-ference contrast microscopy, which because the numerical aperture of the condenser can be fully utilised gives superior resolution to phase-contrast. Further, the image is not degraded by the halo effect. Fine details such as spindle fibres can be resolved in living cells. Another advantage is that normal bright-field objectives can be used.

(6) The sixth method is by interference microscopy, in which the mutually interfering beams which produce the contrast are obtained by an *interferometer system* incorporated in the microscope itself. Again resolution is better than with phase-contrast, since the aperture of the objective is unrestricted by a phase plate and the image undegraded by the halo effect.

Another important application of the interferometer microscope is in the determination of the dry mass of living cells. The thickness and volume of cells, and the concentration of solids and of water can also be deduced in certain instances.

BULK FIXATION

Principles

The purpose of fixation is

(1) *Coagulation* of the constituents of the cell without solution, disintegration, or disturbance of their internal structure or external spacing. The curd must be life-size and it must contain the water, which makes up nine-tenths of the cell.

(2) *Mordanting,* in a broad sense, to produce surface conditions in the fixed structures that will enable them to hold theparticular stains intended for making them visible.

(3) *Toughening* the curd so that it will resist embedding, or *softening* it so that it will not resist squashing.

The objects to be treated have certain invariable properties. The chromosomes are threads composed of fibrous protein. These threads are the most resistant material in the cell, and they are further strengthened during *mitosis* by being spirally coiled and covered with a sheath of nucleic acid. It is this uniform coat which makes the chromosomes stain with the same basic dyes in all organisms. Acetic acid swells them slightly, *chromic* and *osmic acids* contract them, but none of these reagents causes displacements.

The chromosomes themselves, therefore, present little difficulty and little variation in fixation. They exist, however, under various conditions of cell and tissue environment. The cell -environment is of two kinds :

(*a*) The spindle and cytoplasm which, though more watery than the chromosomes, nevertheless support them well under all but the worst conditions of fixation.

(*b*) The prophase nucleus, which is particularly fluid in the prophase of meiosis. Its coagulation without displacement by shrinkage requires rapid fixation and careful handling.

The tissue environment is of greater importance in plants than in animals for here the cell walls, and also hairs and wax.on the epidermal surface, are an obstacle to mass fixation. They demand a special treatment for rapid penetration. It is therefore in relation to the cytoplasm and the cell walls, on the one hand, and the after-treatment, especially the staining, on the other hand, that we have to consider the choice of fixing reagents for chromosomes.

Practice

Coagulation can be brought about by boiling water, dilute sulphuric acid, absolute alcohol, and a hundred other agencies. The boiling water uncoils spiral chromosomes, and either water or alcohol will cause collapse of *prophase nuclei*. Single cells can even be fixed by drying. Smeared pollen mother cells, for example, can be fixed in a *desiccator*. In this way separate,punctured eggs of worms have shown excellent fixation and pollen grains from *herbarium sheets* have given satisfactory pictures of the nuclear conditions.

Coagulation without disturbance can be best brought about by two single reagents : 45% acetic acid as a *fluid* and *osmic acid* as a vapour.

Acetic acid, with its small ions, penetrates a tissue rapidly, but it swells the protoplasm and does not toughen it for later treatment, or at least not quickly enough. This difficulty may be *surmounted* either by combining the stain with the acid for joint action, or by combining alcohol with the acid to fix and harden the protein, other than the chromosomes, in the cells so that the chromosomes may be contained and preserved for later staining.

Osmic acid vapour gives the best possible fixation, but without deep penetration, or indeed any penetration of cellulose walls. It can therefore be used only on animal cells, which should be smeared in a thin film and inverted over a chamber containing the 2% solution for half a minute. Since the preparation must afterwards be stained and therefore subjected to either dehydration or maceration, it must be toughened after fixation. This can be done with 1 % *chromic acid* for 1 hour.

The combination of all these three acids has the necessary properties of *penetration*, *coagulation* and *toughe ning*. It was the foundation of the first effective fluid for fixation in bulk. This combination in weak concentration was first used by Flemming in 1882. ts variations have given rise to all the chief fixatives in use today. These variations have consisted in

(1) Changes of concentration e.g., strong Flemming etc.

(2) Reduction in the acetic component, to preserve chondriosomes from solution, e.g. Benda, or for smear preparations where penetration is less important, *e.g.* La Cour's 2BE. The complete omission of acetic acid by Champy and Lewitsky is unreliable for plants on account of the cell walls, but Minouchi, Matthey and Koller have found it useful for mammals.

(3) Replacement of osmic acid by less expensive reagents. This was done completely by Bouin in 1896 with picric acid for animal material and by S. Navashin in 1910 with *formaldehyde* for plant material. t has been done partially in La Cour's series of fixatives with potassium

dichromate. Picric acid, as well as other substitutions by platinic chloride, ruthenic acid, uranic acid and uranium salts we cannot recommend.

(4) Addition of reagents to reduce surface tension, or to raise the osmotic pressure, as discussed later.

A second line of development arose from the penetrating *alcoholic fixatives* first used by Carnoy. Their disadvantage lies in their unsuitability for embedding afterwards.

The aqueous fixatives harden a tissue for embedding. The alcoholic fixatives containing acetic acid soften cell walls and matrices conveniently for *squashing*, and by their lower surface tension allow of the *penetration* of resistant membrances such as egg shells.

Fixatives containing both oxidising and reducing agents, suchh as the Navashin and Bouin series, must have the opposite components kept separate until they are used, and should be used immediately on mixing. It is also said that the Champy series shotfd'be fresh, on account, perhaps, of its high osmic content.

Mordanting and Hydrolysis

The essential step in the fixation-staining process is in general one of salt formation of a non-specific character, and Feulgen's method of staining, which will be specially considered, gives a solitary example of a known chemical reaction being concerned.

In the salting process combination of stain and fixative is vital. Oxidising agents, as well as strong acids, prepare the surface of chromosomes and other bodies for particular staining reactions. Thus a formalin fixative probably gives a weaker nucleic acid reaction with crystal violet than does an osmic fixative.

This difference shows only in *nucleoli*, the chromosomes being saturated after either fixation. Chromic acid, or some equally penetrating oxidising agent, is a necessary component of any fixative for *crystal violet*. On the other hand it interferes with the action of any *acetic stain*.

With some stains, however, no fixation can provide a sufficient mordant for staining. Mordanting must be carried out afterwards. The function of the mordant is either as an iso-electric point modifier or as a chemical link between the stain and its recipient.

Possibly chromic acid before crystal violet and alum before haematoxylin has the first effect, while hydrolysis before Feulgen staining is known to have the second effect. An after mordant effect is produced by rinsing a slide in iodine or in chromic or picric acids after staining with crystal violet.

Penetration

Rapid penetration of the fixative is necessary for its uniform action. This is aided in six ways

(1) Dissecting out small testes (under Ringer-solution) or anthers, cutting up under the fixative, *slitting root-tips* or *truncating anthers*, or *smearing material* before fixation. In addition

 (*i*) Animal eggs may be pricked to admit fixative.

 (*ii*) Pollen mother cells which hang together and are thus impossible to smear (e.g. in some *lilies* and *cereals*) can be squeezed out of cut anthers in four strings and fixed naked for embedding.

(2) Penetration of the surface by *alcoholic fixatives* is most satisfactory, but they fail to harden the important cell structures. To remedy this they can be followed after one minute by rinsing and the use of an aqueous fixative (*Kihara's Carnoy-Flemming method*). This method is useful where flower buds have to be fixed whole.

(3) Some fixatives, *e.g.* Allen's Bouin or acetic acid, can be used warm (36°-40°C). This increases the rapidity of penetration and it may increase the effectiveness of coagulation and hardening.

(4) After unheated aqueous fixation (but never after alcoholic treatment) the material should immediately be shaken and put under a water or hand vacuum pump to remove air bubbles which may delay penetration.

(5) Materials may be added to an aqueous fixative to lower its surface tension. This is the purpose of saponin in La Cour's fixatives.

(6) Maltose and urea have been used to adjust the osmotic pressure of fixation. The 'effective' osmotic pressure is, however, that due to small ions in solution and these substances are relatively useless. Adjustment of the effective osmotic pressure to that of the tissue to be fixed is important in bulk fixations of animal material and can be achieved by making up the fixing solutions in normal saline.

Condition of the Material

Material before fixation must be healthy. Starvation with.animals, *drought* with plants, have a disastrous effect on the fixation properties of both chromosomes and *mitotic spindle*. *Root tips* for mitosis are best either taken from pot plants where the ball of soil is neither too wet nor too dry, or from *water-cultures*. Testes for meiosis should usually be from immature or young animals. Cold treatment is a *palliative* for poor condition in hot weather, and even the use of fixatives at freezing point.

Duration

Owing to the absence of cellulose walls, animal tissues are fixed more rapidly than those of plants. The hardening process, in both cases, requires several hours. The upper limit varies with the reagent as follows

Fixative Type	*Limit of Safe Action*	*Effect of Excess*
Formalin	several weeks	—
Low Osmic	1 week	—
Alcoholic	24 hours	reveals spindle structure
High Osmic	1 hour	damages chromosomes

Since animal cells are not hardened enough in one hour, their fixation in *high-omic* types must be followed by 12-24 hours in 1% *chromic acid.*

Uses of Bulk Fixation

Bulk fixation was formerly used for preparing the material for embedding and sectioning by the *paraffin method.* Now it can also be used for the direct preparation of squashes.

SMEARS AND SQUASHES

Material

Sections have now been largely replaced by smears and squashes for. all but the smallest masses of material. The advantages of these methods are rapidity of fixation and rapidity of handling without embedding. In both respects rapidity means not only saving of time but also increased efficiency. Both methods enable us to examine single layers of large cells in their totality. This advantage is combined with the instantaneous fixation of smears and the rapid acetic fixation of squashes.

Smear Methods

Where feasible, smearing is to be preferred to squashing. for meiosis in anthers and soft testes. Smearing consists always (in the absence of pretreatment) of the direct spreading of the cells in such a semi-fluid tissue over the surface of the slide with a *flat-honed* scalpel and the immediate inversion of the slide over a dish of fixative. The combinations of fixative and staining that are possible with smearing are shown in. Schedule 2.

Acetic Squash Methods

The simplest of all chromosome treatments is to use a combined stain-fixative of the aceto-carmine type first used by Schneider and adapted for chromosome work by Belling in 1921. For this purpose inversion is unnecessary, since the cells can be directly mounted and studied in the fixative. This makes acetic staining methods very convenient for determining. stages of division in anthers and testes before fixing them by some other method.

The standard iron-aceto-carmine method was devised by, Belling for the study of pollen mother cells, but it is now equally widely used for animals. It consists in *teasing out* the tissue in *aceto-carmine* with unplated iron needles and mounting directly under a cover slip. The iron acetate formed acts as a mordant for the carmine and gives a differentiation of chromosomes and *cytoplasm* which is usually good enough for the study of *metaphases*.

In many flowering plants, however, particularly where the chromosomes are small, the cytoplasm takes up too much of the stain.. The following improvements and developments of the aceto-carmine technique are chiefly concerned with this difficulty.

(1) *Fixation.* A prior fixation in 1 : 3 fresh acetic alcohol for 12-24 hours reduces the staining of the cytoplasm. Whether this effect is due to differential solution or change of pH is not known. Propionic acid can be used to replaceaceticc acid with advantage for some materials.

(2) *Storage of Material.* If the material is now wanted for immediate staining, it can be stored in 70% alcohol at 0-4°C, but fixation and staining will suffer after two months or more. Long storage however can be used for the special purpose of exaggerating the differentiation of the spindle to facilitate the study of its development.

(3) *Spreading.* Gentle heating of the slide over a spirit flame flattens the cells, sticks them to the slide and cover slip and spreads the chromosomes whether in prophase or metaphase. The degree of spreading is due to the pressure of the cover slip, and this is controlled partly by the amount of material on the slide, partly by the amount of stain-fixative used and partly by pressure on the cover slip. All useless debris should be removed, preferably with the aid of a dissecting microscope.

This is essential for the flattening and spreading of the chromosomes at pachtene. Gentle finger tip pressure on the cover slip under filter paper may assist in this process, but lateral movement of the cover slip should be avoided if entire cells, are required. A minimum-sized drop of stain-fixative, compatible with allowing the placing of the cover slip in position without air bubbles, should be used. To obtain even staining, 30 to 60 seconds should elapse before applying the cover slip.

(4) *Pre- and Post-mordanting.* The method of Thomas (1940) is useful for small chromosomes. Iron acetate is introduced, into the acetic alcohol. A weak stain is best, about 1/3 strength *aceto-carmine* (diluted with 45% acetic acid) and more iron is introduced at the time of teasing by using steel needles.

Alternatively a trace of *ferric chloride* may be added to the acetic alcohol (a few drops of concentrated aqueous solution) to give the fluid a pale straw colour. This method leaves the cytoplasm in a soft and plastic condition facilitating easy flattening.

Iron alum can be used as a mordant after fixation. It is a most satisfactory method for Algae and for pachytene in the tomato when followed by differentiation in 45% acetic acid over a steam bath. Prolonged staining in *alcoholic hydrochloric acidcarmine* may prove advantageous for some material.

(5) *Acetic-Orcein.* This chromosome stain recommended.by La Cour is more selective than carmine and for many organisms gives clearer permanent preparations. No iron mordant is needed. It can also be used in 45% propionic acid with advantage for some species.

It is now the recognised stain for salivary gland chromos somes of Diptera.

Acetic-oreein has also proved valuable for the chromosomes of mosses. In some tissues and organisms, the selectivity of staining can be further improved if, after acetic alcohol fixation, the tissues are hydrolysed for 5-8 min in N HC1 at 60°C before staining. This has proved advantageous in various species of Ascomycetes.

(6) *Lacmoid Indicator* (resorcin blue). This reagent was first used in alkaline solution for staining cytoplasm (v. pollen tube technique). In acetic acid solution it stains only the chromosomes.

Again no iron mordant is needed. Its colour remains red so long as the preparation remains acid and the mount is made in cedarwood oil. If the mount is made in neutral balsam the colour changes to blue. Lacmoid is most useful after maceration.

(7) *Bismarck Brown.* Formerly used as an aqueous stain this reagent can likewise be substituted for carmine.

(8) *Compound Staining.* Nebel has suggested the use of chlorazol black E as a combined stain with carmine for small chromosomes. Workable stains can also be made from all[combinations of orcein, lacmoid and carmine.

(9) *Storage of Slides.* Preparations are usually improved after keeping for a few days. The edges of the cover slip should be sealed with a rubber solution, or a mixture of gelatin and acetic acid. They are best kept in a refrigerator at 0-4°C.

(10) *Permanent Method.* There are a number of methods for making permanent preparations, the most notable being McClintock's, the vapour method of Bridges and the dry ice method of Conger and Fairchild. The last method should be avoided with carmine, for it is easier to avoid change, or correct for over-staining with other methods. A modified form of McClintock's method is given in the schedule.

Combined fixing, staining and mounting methods developed by Zirkle can be used with water-soluble or fat-soluble mounting media. The results are rapid and often satisfactory. Orcein and lacmoid can be used to replace the carmine in any of these formulae.

In conclusion we may say that no one type or method of acetic staining-fixation is best for all materials. On the whole orcein and lacmoid are better than carmine for all animal tissues. Both are excellent also for many plants, but may burst some pollen mother cells whose cytoplasm is delicate, *e.g. Allium.* Owing to its slight hardening effect, for pollen mother cells of plants with small chromosomes and dense cytoplasm, *e.g. Rosaceae,* aceto-carmine combined with premordanting in the most satisfactory. For some organisms and special tissues orcein is better than either *carmine* or *lacmoid.*

Maceration

Soft fixations of soft tissues, i.e. alcoholic fixations of PMC -or insect testes, need not, indeed should not, be macerated for squashing. All other material requires maceration, which must be so adjusted in plant tissues that solution of the *pectic salt* in the middle *lamella* of the cell walls is complete without the cell contents being unduly softened. There are eight methods of maceration

(1) Add N HCl to acetic orcein or acetic lacmoid.

(2) Extend Feulgen hydrolysis to maximum time.

(3) Immerse in equal parts of conc. HCI and 95% alcohol for 5 min without warming. Further, after alcoholic fixation, harden the material in Carnoy for 10 min before staining.

(4) Fix in 5 N HCI for 15 min at 20°C.

(5) Treat for 2-5 min in sat. aqu. ammonium oxalate and H_2O, (equal vols of each) but avoid tap-water for rinsing. Use when the fixative contains osmic acid (Ford unpub.). For meristematic tissues only.

(6) To assist maceration after osmic fixation, fix for 2 hr and then treat tissues overnight in equal parts 1 % chromic acid and used fixative. If required the treatment can be followed by that in (5).

(7) Treat with snail stomach cytase overnight or 5% pectinase in 1 % aqu. soln. peptone 4-16 hr. Use for all plant tissues except pollen grains.

(8) Treat PMC with large chromosomes for carmine staining with a fungal enzyme, clarase, 1% soln. for 10 min to 3 hr.

Root Tips. All these methods except the last are suitable. There are some difficulties however in plants with numerous small chromosomes. Most of the metaphase plates are then found to be in side view, and the cells cannot be turned without damage. Pre-treatment with drugs can overcome this difficulty by inhibition of the spindle; metaphase plates with super-contracted chromosomes are then accumulated in the tissue.

To avoid too much contraction O'Mara immerses excised roots in 0.01 % colcbicine for less than 3 hours. As an alternative a-bromonaphthalene in saturated solution is cheaper and as satisfactory for most species. Low temperature can produce a like effect. Thus transference of *Trillium grandiflorum* from 24°C to 0°C for two days greatly increases the proportion of.mitoses which are in metaphase without causing too much contraction.

Warmke's method of keeping cut root-tips for 2 hours at 0°C before fixation probably serves the same purpose. These treatments can be followed by *squash* methods, but where super-contraction must be avoided, one of the following two methods may be tried

(*i*) (*a*) Carnoy fixation. (*b*) Transverse sectioning by the freezing microtome. (c) Combined softening and staining in acetic-lacmoid.

(*ii*) (a) Aqueous fixation. (b) Transverse sectioning by the paraffin method. (c) Feulgen staining with extended hydrolysis. (d) Pressure applied after mounting in thin balsam.

Leaves. Where roots are unobtainable rapidly growing leaf buds (the leaf meristem) as well as petals, tendrils or glumes provide a valuable source of mitoses. Again to accumulate metaphases treatment of excised leaf tips in 0.2% colchicine, kept in light for 1-2 hours, is suggested by Meyer. We find that the concentration of colchicine can be as low as 0.05% depending on the greenness of the leaves.

Pollen-Grains. Method (3) is the most suitable. Maceration allows the contents of the pollen-grain, whether large as in *Lilium* or small as in *Pyrus,* to escape from the thick wall which fn these cases stands in the light of the chromosomes. An exine solvent has been used on *Gossyptum* pollen by Bernardo.

Embryo-Sacs. Method (1) has a sufficiently hardening effect on the cytoplasm to preserve the embryo-sac entire.

PARAFFIN METHODS

Value

Embedding may become necessary where, for one of several reasons, smears or squashes are impracticable. This is particularly so in the handling of very small organs, including all but the largest embryo-sacs, or even for larger bodies where mitoses are scarce. As an auxiliary method, embedding is also useful in showing the arrangement of cells in a tissue and the sequence of the stages of meiosis in a testis or an anther.

Fixation and Washing

For material that is to be embedded, aqueous fixatives are necessary to protect the cells from maceration by water or collapse by loss of water. The precautions already described must be taken to ensure satisfactory penetration.

A solution of at least 20 times the volume of the material to be fixed must be used to fix it, and bulk material in any case requires stronger fixatives than do smears.

For most purposes long washing in water is unnecessary and even harmful. Indeed washing can be omitted altogether in preparing root tips for chromosome counts. The root tips can be taken directly from the fixative into 70% alcohol. For more delicate tissues 2 or 3 changes of water are preferable before transference to alcohol.

Dehydration and Infiltration

Many attempts have been made to find reagents as suitable substitutes for ethyl alcohol and xylol, or ethyl alcohol and chloroform. The objects have been a speeding up of theprocess, an avoidance of excessive hardening and shrinkage of tissues, and the use of one fluid instead of two. The most important of these reagents are

(1) *Dioxan* : Graupner and Weisberger; Johansen; Baird (animal tissues); La Cour; Maheshwari.

(2) *n-Butyl alcohol* : Zirkle, Randolph.

(3) *Tertiary butyl alcohal* : Johansen.

(4) *Isopropyl alcohol* : Bradbury.

(5) *Methylal paraffin oil* : Dufrenoy.

Of these reagents, the first three have given the most promising results. Dioxan, however, has toxic properties and extreme case should be taken in its use. Johansen claims that his method is the least drastic and does not remove all the bound water from the tissues. It should therefore avoid collapse of delicate structures such as prophase chromosomes.

All substitutes for chloroform are nevertheless poorer solvents of paraffin wax and for the infiltration of bulky material, such as flower buds, chloroform is to be preferred. Recent improvements, however, in the quality of ester waxes now make it possible to avoid the reagents largely responsible for excessive hardening and shrinkage.

Ester wax will tolerate up to 5% of water in solvents such as ethyl alcohol so tissues can be infiltrated with the wax direct from 95% alcohol if desired. Special treatments are useful for small objects. Eosin may be introduced in the 70% alcohol, or fuchsin in the 1 : 3 alcohol-chloroform, to make them conspicuous in the wax.

Embedding

The following precautions should be borne in mind

(1) Remove all traces of the solvent by evaporation, or by changing the molten wax.

(2) Avoid overheating. The oven temperature should never be more than 60°C.

(3) Choose wax of melting point suited to the tissues to be cut and the laboratory conditions. In a temperate climate use 50°C m.p. wax for sections to be cut at 14-40 μm and 58°C m.p. below 14μm. Waterman has described the preparation of hardened paraffin waxes having low melting points; ester waxes with these qualities can be obtained commercially.

(4) Orientate root tips and anthers in rows in the wax for convenience in cutting. Randolph has suggested a method of card mounting which might be valuable for small root tips. The tips are mounted prior to fixation. Faberge and La Cour have devised an electrically heated needle to facilitate orientation, avoiding the danger of damaging delicate tissues by overheating.

(5) Cool the block rapidly in water to avoid crystallisation of the wax.

Section Cutting

The first question to decide is the thickness at which the sections should be cut. Until recently, for heavy haematoxylin staining, sections had to be cut so thin that a large proportion of nuclei were injured. Indeed some microtomes are not adjusted to cut at more than 20 μm. The best extend to 40 μm. The following list shows how to adjust the thickness of the sections to the length of the chromosomes:

Length of Chromosomes		1-10 μm	15-40μm
	RT	4-16μm	20-40 μm
Thickness	PMC	14-20 μm	30-40 μm
of	EMC	20-30 μm	40 μm
Sections	Testes and Eggs	4-16μm	20-40/Am

(1) Keep the microtome knife sharp by correct honing and stropping. For a review of methods.

(2) To secure straight ribbons with a minimum of breaking, trim the block so that the wax is evenly distributed around the material. The sides must be parallel and mounted parallel to the razor edge.

(3) Mount the razor at the correct angle and screw it tight to keep it rigid.

(4) If there is difficulty in securing continuous ribbons of thick sections in cold weather, either warm the mounted block and holder in the oven, or warm the edge nearest the razor with a hot scalpel.

(5) For thin sections that wrinkle badly, owing to the wax having too low a melting point for the room temperature, cool the block on ice or use a hardened paraffin.

(6) To secure sections of hard material (animal eggs) expose one side of the material by trimming the block, and soak in water 12-24 h. Waddington and Kriebel advised the use of a small

amount of petroleum ceresin in the embedding wax.

(7) In dry climates the ribbons will be electrified. Hance, has suggested air conditioning to avoid this. For other methods.

Mounting Ribbons

Mayer's albumen is most commonly used for fixing ribbons to the slide. The film must be the thinnest obtainable by smearing with the finger. Even so, it will be well to avoid loss of thick sections through failure of drying, by heating the smeared slide gently over a spirit flame for 2-3 seconds.

The ribbons are cut into lengths shorter than the cover slip to allow for stretching.

The ribbons are floated on a few drops of water on the slide. A hot plate maintained at 45°C is necessary for stretching and drying the ribbons. Avoid air bubbles under the sections. "They arise from careless laying out or too sudden beating. Drying is completed in 4 to 12 hours according to thickness. It,can be hastened by the use of 20% alcohol instead of water.

Removing the Wax

Before the ribbons can be stained the wax must be dissolved in xylol and the slide then placed in absolute alcohol and after bleaching if necessary taken down to water.

An excellent review of sectioning methods has been written by Steedman.

Note. In Britain permits for duty-free absolute alcohol are obtained from a Customs and Excise Office.

STAINING AND MOUNTING

Staining

Apart from the acetic stains which act as combined stainfixatives, nine basic dyes are suitable for aqueous staining of chromosomes, as follows

(1) The *leuco-basic fuchsin* method was first developed by Feulgen and Rossenbeck as a microchemical test for the desoxyribosc-nucleic acid (DNA) found in chromosomes. It depends on Schiff's aldehyde reaction, and gives a translucentt but fairly permanent stain. The method consists of

(*i*) Liberation of the aldehyde groups of the nucleic acid by mild hydrolysis in normal hydrochloric acid at 60°C.

(*ii*) Chemical reaction between the liberated aldehydes and the leuco-basic fuchsin, resulting in a violet colouration of the chromosomes.

Three types of substance in the cell are capable of giving the aldehyde reaction. (*i*) free *aldehydes* in lignified cell walls, (*ii*). DNA after hydrolysis, (*iii*) *polysaccharides* after oxidation with chromic acid : (*i*) and (*iii*) thus require no hydrolysis but (i) is confined to the cell wall and (*iii*) is limited to the effects of chromic acid fixation. A specific diagnosis of DNA acid in the cell should therefore always be possible by means of the Feulgen reaction.

Indeed the concept of DNA constancy for the haploid chromosome set is partly based on the quantitative estimation of Feulgen values in nuclei by microphotometry. It should be appreciated, however, that probably only a relatively constant fraction of the DNA in the chromosomes reacts to the Feulgen stain.

The method is of wide application. Tissues can be stained in bulk (see squash methods), or as sections or smears. It can be used after alcoholic fixation, and unlike most chromosome stains, it requires no differentiation.

The Feulgen method is subject to occasional failure or apparent failure. This failure or seemingly weak reactivity has been due to various causes

(*i*) Diffuse distribution of DNA in extended chromosomes or expanded nucleus, e.g. lampbrush chromosomes or the vegetative nucleus in pollen grains.

(*ii*) Incorrect timing of hydrolysis. As to the minimum and optimum times of hydrolysis, Hillary, using nucleic acid impregnated agar blocks and four different types of fixatives,. has shown that two types of hydrolysis curve are possible.

These depend on the presence or absence of chromic acid in the fixative. With fixatives containing chromic acid, maximum staining follows hydrolysis at 60°C from 6 to 30 min; without -chromic acid the maximum extends only from 4 to 8 min. After storage of tissues in alcohol it is often necessary to extend hydrolysis times. Further, hydrolysis sometimes has to be extended, possibly because in some plants the tissues contain substances inhibiting it, *e.g.* in root tips of *Chrysanthemum*. In some instances this may perhaps be due to interference by protein.

(*iii*) Osmic acid unless carefully reduced by bleaching will sometimes interfere with the Feulgen reaction. This is particularly serious with small chromosomes. Incomplete bleaching of the nucleolus has its use, however.

(*iv*) Formalin fixatives will lead to staining of the cytoplasm by the Feulgen reaction, if not washed out.

(*v*) Poor samples of basic fuchsin, which fail to give completely decolourised solutions, will lead to stained cytoplasm, and sometimes weakly stained chromosomes. Coleman has suggested decolourising carbon as a means of obtaining colourless solutions free from impurity.

(*vi*) Kasten has listed 28 other dyes of various colours which will give a Feulgen-type reaction following acid hydrolysis. By choosing dyes of contrasting colours, it is possible to intermingle cells from different treatments in the same squash preparation, in order to obtain comparative measurements in equally squashed cells.

(2) The *crystal violet* method is simple and rapid. It is possible to obtain sections up to 40 μm in thickness with wellstained chromosomes and clear cytoplasm. Also the intensity of the stain is easily reduced by differentiation, an important factor in the study of large chromosomes at meiosis. A weak stain, 0.1%, recommended by Upcott and La Cour, can be used for this purpose. Where the stain does not hold in small chromosomes the chromic acid method of La Cour or the picric acid method of Johansen can be used.

(3) *Haematoxylin is* now seldom used except where intense permanent staining is required. Particularly in plants, colouration of the cytoplasm is difficult to avoid, and this stain is therefore more easily applied to thin sections, or smears, of animal tissues. A more elaborate method than that given in Schedule 10 is recommended by Cole.

(4) The *brazllin* method of Belling and Capinpin. suffers from the same defects as haematoxylin, but it can be made to yield good results with pollen mother cells.

(5) *Celestln blue B* and *gallocyanin* can be used as simplenuclear stains, preferably as iron lakes which can be prepared by boiling *2-3* min in a 5% solution of iron alum. No mordanting or differentiation is required.

(6) *Giemsa* (Gelei's modification) has been recommended by Belar as a satisfactory stain for animal chromosomes after fixation by osmic vapour. Giemsa has also been used for staining mitosis in germinating spores of bacteria after fixation for 3-3 min in osmic vapour and for an indefinite time in 70% alcohol followed by hydrolysis for 7 min in N HC1 at 60°C.

(7) *Azure A* in the presence of SO, stains nuclei selectively after a short hydrolysis in N HC1 at 60°C. Thionyl chloride has been used by De Lamater as a source of SO_2 in the use of this stain.

(8) *Toluidine blue* is an effective stain for chromosomeswhen used following hydrolysis in N HCl at 60°C, or 5 N HCI for 40 min at 20°C to remove ribose nucleic acid (RNA). When, combined with dimetbyl hydantoin formaldehyde resin, it provides an effective stain-mounting medium, particularly for spreads of mammalian chromosomes.

It is most useful in autoradiography (stripping-film technique) for staining through, the film after photographic processing. Stain for 30-60 seconds in a 0.05% aqueous solution. Used hot it is possible to stain nuclei, nucleoli and chromosomes effectively in 'Epon' resin sections as thin as 65-70 nm.

(9) *Methyl green* as a nuclear stain is perhaps best known for its use in combination with pyronin. As used by Kurnick it is believed to be specific for polymerised DNA, but this has. been questioned by Alfert (1952). It is an unstable dye and solutions have to be repeatedly extracted with chloroform before use, to remove methyl violet.

Light green and fast green are suitable counter stains to use for the nucleoli after the chromosomes have been stained by the Feulgen method. The light-green method of Semmens and Bhaduri gives specific staining. A simple fast green method for slide or bulk material (0.1% aqueous solution for 2-12 hours) has been recommended by Hillary to be used after bleaching in SO_2 water and rinsing.

Table7.1: Refractive Indices (n) of Mounting Media, etc.

Medium	*n*	*Medium*	*n*
Distilled water	1.336	Crown glass	1.518
Sea water	1.343	Cedarwood oil. thick	1.520
Liquid paraffin	1.471	Balsam in xylol	1.524
Olive oil	1.473	Clove oil	1.533
Glycerine, concentrated	1.473	Balsam, dry	1.535
'Euparal'	1.483	Gum damar	1.542
Xylol	1.497	'Clarite X'	1 567
Cedarwood oil, thin	1.510	(== 'Nevillite No. 1')	

Moulting

Optically the soundest mounting medium is one whose refractive index lies between that of the glass slide and that of the tissues, which may be a little higher. Canada balsam in general has the advantage on this ground. It is however acid, and causes certain stains to fade. To overcome this difficulty various neutral resins have been produced. Neutral balsam in sealed tubes and 'clarite X' may be recommended, but clarite cannot be used for acetocarmine preparations, and becomes milky with alcohol.

Xylol is the usual solvent for such resins, but toluene dries more quickly. Soft tissues or squashes are liable to collapse in these petroleum distillates. To avoid this we may avail ourselves of two other methods

(1) *Feulgen, carmine or orcein preparations :* Mount directly from absolute alcohol in 'Euparal' or balsam dissolved in alcohol. If the alcohol contains any water the balsam will become cloudy. Warm on the hot plate to get rid of this. Dioxan and dioxan-balsam are recommended by Hillary.

(2) *Lacmoid preparations :* Mount directly from absolute alcohol to

Fading

All stains can fade in certain circumstances, though what these circumstances are has not always been defined. The chief agents of fading seem to be (*i*) ultraviolet light (since it will happen rapidly from an arc lamp projector and gradually from daylight), (*ii*) acidity of the mounting medium, already referred to, and *(iii)* inclusion of reagents from earlier treatment, through careless mounting.

From all these causes crystal violet fades most easily (even in a few weeks), haematoxylin and carmine least easily. F. H. Smith claims that washing in saturated picric acid in absolute alcohol after staining makes crystal violet permanent. His method certainly delays fading. The chromic acid aftermordant would probably have the same effect.

Re-staining

After fading or unsuccessful staining re-staining is possible with the same stain or with any other stain for which the original fixative was appropriate, *e.g.* a Flemming crystal violet preparation can be re-stained in Feulgen. This is done by repeating the process after removing the cover slip and the mounting medium with xylol.

To this rule there is one physical exception : an acetic smear is likely to collapse on removing the cover slip. And there is one chemical exception : Feulgen preparations cannot be re-hydrolysed for re-staining with leucobasic fuchsin.

They can however be re-stained, after acetic fixatives with an acetic stain (especially lacmoid), after aqueous fixatives with crystal violet and after alcoholic, with haematoxylin. Further, after acetic fixatives an ordinary Feulgen preparation which has turned out to be insufficiently stained may at once be brightened with orcein or carmine (2-5 min).

Bleaching

Before staining, sections and smears should be bleached in hydrogen peroxide where necessary. Bleaching is required to oxidise the black precipitate produced by fats with fixatives containing osmic acid. With bulk material for Feulgen squashes bleaching is usually not necessary. Hydrolysis removes part of the precipitate and squashing makes the remainder negligible.

SPECIAL TREATMENTS

Pre-Treatment for Structure

Internal spirals, which appeared at metaphase by accidents of fixation, were first illustrated in *Tradescantia* by Baranetzky. Relic spirals have long been known in the resting and prophase nucleus. For a long time a relationship of the two was vaguely presumed. The general laws of chromosome structure were clarified only by the regular demonstration of internal spirals by pre-treatment of pollen mother cells (or testes in the few animals examined).

This method derives from Sakamura who photographed major spirals in MI chromosomes of *Tradescantia* fixed in boiling water and unstained. A second step was the discovery by various workers that smears dried before chromic fixation (accidentally or otherwise) showed spiral structure. Diploid and tetraploid species of *Tradescantia* and many species of *Trillium* are convenient for the smear treatment of PMC necessary with all methods.

Some of the methods, such as heat and ammonia vapour, depend on uncoiling and can restore a resting nucleus condition. Others possibly depend on dissolving the histone associated with the DNA backbone.

A third effect is swelling., which depends on increased hydration following a change in pH of the cytoplasm away from the IEP of the chromosomes.

(1) *Meiosis*

(1) *Ammonia vapour* (PMC in 3% sucrose : 5-15 seconds before fixation).

(a) Aceto-carmine.

(b) 2BE-crystal violet or Feulgen.

(*ii*) *Ammonia in* 30% *alcohol* (6 drops in 50 ml. 5—20 seconds). Flemming-crystal violet. Useful because alcohol attaches the cells to the slide.

(*iii*) *Weak alkaline solvents of nucleic acid.* Smear is placed in solutions of NaCN, $NaHCO_3$ or even 10^{-2} M NaOH for 15 seconds to

3 min according to the material. Flemming with CV or Feulgen.

(*iv*) *Acid fumes* (nitric, hydrochloric and acetic) followed by 2BE-CV.

(*v*) *Precipitation of desoxyribose-nucleic acid and partial digestion of protein :* Hillary's modification, 1940, of Casperssion's. method.

(*a*) Pre-treatment as in (*iii*), 30 seconds in 10^{-2} M NaCN.

(*b*) Fixation in Flemming.

(*c*) Preparation left about 12 hours in 0.1 % lanthanum acetate to precipitate nucleic acid as lanthanum salt.

(*d*) 24 hours at 37°C in 1 % solution of trypsin containing a trace of lanthanum acetate.

(*e*) Stain by Feulgen.

(2) *Mitosis*

Most methods of pre-treatment are usually unsatisfactory for chromosomes within a cell wall. As a first attempt an ammonia-thorium nitrate method was suggested for root tips by Nebel. Later Coleman was successful with treatment of smears of pre-meiotic mitosis using a solution of 10^{-5} M $NaCN_2$. Three other methods may sometimes overcome this difficulty

(*i*) Pollen tube mitosis. Aceto-carmine, penetration is easy on account of extremely thin cell wall.

(*ii*) Pollen grain, first mitosis. Pressure in fixation with aceto-carmine.

(*iii*) Root tip frozen, while growing, for four days to give differential reactivity of heterochromatic segments, which then show their spiral structure.

For revealing spiral structure in the chromosomes of cultured human leucocytes, Ohnuki (1968) devised a special hypotonic solution consisting of equimolar (0.55 M) solutions of KCL, $NaNO_3$ and CH_3OONa in the proportion of 4 : 2 : 0.8, respectively. The cells, as a loose pellet, are re-suspended in the hypotonic solution and treated for 90 min at 21-23°C.

Finally micro-incineration has been used as a method of showing chromosome structure.

Micro-Chemical Tests

The effect of DN-ase on lampbrush chromosomes has shown, that DNA runs continuously throughout the chromosome axis. Histone is bound to DNA by salt-like linkages and is situated in the groove of the double-helix. The fission of the chromosome axis with DN-ase indicates that other protein is attached laterally.

Proteins

A number of relatively simple histochemical methods for the demonstration of proteins in cells are available. Chromosomes of plant and animal cells have been shown to react more strongly than the cytoplasm but not differentially. Recently the Sakaguchi reaction for arginine has been modified with the view that quantitative measurements might be mademicrophotometrically.

Staining of tissues in fast green at pH 8 after removal of nucleic acids has been used for the estimation of basic protein in nuclei. It has recently been used by Bloch, in combination with eosin to stain histones in animal tissues.

Nucleic Acids

Two types of nucleic acid are found in cells (*i*) DNA, staining with Feulgen's reagent is present in the chromosomes, nucleolus, mitochondria and chloroplasts, and *(ii)* the RNA not staining with Feulgen consisting of three chemically distinct classes formed on DNA templates. The greatest concentrations occur in the cytoplasm particularly through movement from the nucleus as a prerequisite to protein synthesis. Some RNA appears to be present in the chromosomes throughout the whole mitotic cycle. The RNA present in the metaphase chromosomes may be derived largely from the nucleolus. The distribution of RNA in the cell largely governs, and is governed by, its activity and development, particularly in relation to protein synthesis.

(*i*) *DNA*. Some measure of nucleic acid concentrations in cells can be obtained by staining methods. The Feulgen method is valuable for comparative evaluation of DNA content in nuclei within tissues and possibly between related species by microphotometry. Precautions must betaken to ensure the optimum reaction and to rigorously standardise staining conditions.

Recently McLeish and Sunderland have shown that the mean values obtained with Feulgen stain do not differ significantly from estimations of DNA per cell obtained biochemically. The quantitative aspects of microphotometry have been discussed by Swift, Vendrely and Vendrely. Swift and Rasch, Leuchtenberger. A photo-electric scanning device for Feulgen measurements, which is used in combination with a special cell crushing technique, has been described by Deeley.

The availability of actinomycin in tritiated form with high specific activity provides a sensitive means of detecting DNA autoradiographically in fixed cytological preparations in amounts too small to be shown by Feulgen staining. It is applied to sections, squashes or smears like a stain. It has been established that the drug binds selectively to guanine. Its binding to DNA can, therefore, give an indication of the distribution of guanine-rich DNA in the chromosomes and thus a means of detecting one of two main classes of heterochromatin.

Formaldehyde or glutaraldehyde fixation should not be used, because the resulting cross linking of protein to DNA inhibits binding of the drug to the guanine.

(*ii*) *RNA*. A method devised by Brachet can be used for the detection and relative estimation of RNA. Its validity depends on the use of RNA-nuclease (R N-ase), in combination with Unna's methyl-green-pyronin stain. Cell constituents stain pink to red according to the concentration of RNA.

The chromatin stains green or purplish according to the concentration of DNA and RNA. The nuclease-treated control should show no evidence of pyronin staining. The specificity of R N-ase has been examined by McDonald and Kaufmann.

The activity of R N-ase is impaired by osmic and chromic acid types of fixation. Tissues should therefore be fixed in acetic alcohol or Carnoy. Air drying followed by hardening in 95% alcohol or methyl alcohol for 2 hours is sufficient for some types of cell, e.g. yeasts and blood.

Squashes of embryo sac mother-cells and root tips can be subjected to Brachet's test if enzymatic methods are used for cell separation. Fixatives that preserve phospholipids in the cell seem to impair the differential qualities of Unna's stain. Improvements in the preparation of Unna's stain have been made by Taft and Kurnick.

Azure B at pH 4 stains RNA purple and DNA blue-green. With appropriate nuclease extraction of sections, smears or squashes, it provides a useful control in tests for the specificity of radioactive nucleic acid precursors in autoradiography. It has been used for relative estimation of RNA following removal of DNA with DN-ase.

Phospholipids

Evidence for the presence of these substances in chromosomes has been obtained by histochemical and biochemical methods. (*i*) Tissues are fixed in Lewitsky's fluid and the sections stained by a modified acid haematin reaction. The euchromatic parts of chromosomes give a blue-black reaction for phospholipids from mid-prophase to end of anaphase in mitosis and meiosis, the heterochromatic regions give the same reaction at all times except early prophase. (*ii*) Yellow staining of pbospholipid-containing regions is obtained when the sections are stained in a mixture of orange G-aniline blue at pH from 2-3. *(iii)* Methods for revealing the presence of phospholipids histochemically masked by protein interference have been suggested by Serra, *cf.* La Cour *et. al.* and Berenbaum. The acid haematin procedure of Baker is most suitable for showing phospholipids in mitochondria.

For purposes of chemical tests of nucleic acids and proteins in animal cells, nuclei, chromosome threads, and such cytoplasmic particles as mitochondria can be separated with a highspeed centrifuge or by gentle homogenisation in a Waring blender.

It is sometimes necessary to remove one or both nucleic acids from tissues for further histochemical procedures or for investigations in autoradiography. A choice of method is available

(*i*) Digestion of RNA or DNA by specific enzymes.

(ii) Removal of both nucleic acids by treatment of sections or tissues in 4⁰/ trichloroacetic acid at 90°C for 15 min or in 5% perchloric acid at 60°C for 30 min.

(*iii*) Removal of RNA by treatment of sections or smears, after fixation in acetic alcohol or Carnoy, in N HCl at 60°C for 10 min or in 10% perchloric acid at 4°C for 18 hours. The last method is not always satisfactory for plant material.

The theory and application of various histochemical tests are discussed by Pearse.

With the help of these tests we can make a number of clear definitions. *Chromatin* is part of a chromosome thread to which the characteristic DNA is attached. A *chromosome* is a whole thread to which this acid is attached and having the characteristic capacities of spiralisation and replication in a regulated cycle.

A *nucleolus* is a body within a nucleus consisting largely of gene product resulting from the activity of a specific loopforming segment of a nucleolar chromosome. Plate XXIX. Free nucleoli, as in oocytes of amphibia, each contain a DNA replicate synthesised at the same locus a; the attached nucleolus *Heterochromatin* is a part of the chromosome which is out of step with the major bulk of *euchromatin* and replicates late in the cell cycle.

Heterocbromatin, unlike euchromatin, may remain tightly spiralised at telophase and throughout interphase or as found in *Fritillaria lanceolata* remain spiralised at a constant intermediate level for the whole mitotic cycle and also at meiosis from leptotene to zygotene. In animals, particularly mammals, *heterochromatin* is often associated with the centromere or with the nucleolus as nucleolus associated chromatin.

The strong probability that heterochromatin exists in two main classes, with perhaps intermediate variants in respect of spiralisation, is favoured by recent studies on satellite DNA inasmuch as these suggest that the DNA in non-nucleolar heterochromatin consists of repetitive base sequences, either AT or GC in constitution. Finally, with the aid of new techniques now available for recognition of heterochromatin, there should be less confusion with the phenomenon of negative and positive heteropycnosis.

Animal Eggs

Three stages in the development of the egg concern us

(*a*) Prophase of meiosis this begins in the young ovary and may last the greater part of its life.

(*b*) The meiotic divisions : these are rapid and are usually set in motion by the entry of the sperm.

(*c*) The cleavage division of the fertilised egg : these exceedd in their rapidity all other mitoses and in consequence have a character of their own.

During the oogonial divisions and in early prophase neither yolk nor shell is present to hinder treatment, and the ovary can be treated like a testis. Where the shell is thin and the yolk deficient, squash preparations can still be made in all the later stages.

This is true of the Annelida and Mollusca. For example in *Allolobophora* eggs, Foot and Strobell obtained their best preparations of meiosis by puncturing each egg in a very small drop of water. When the contents of the egg were dragged out of its membrane they dried so rapidly as to give admirable fixation of the nucleus in a self-flattened smear.

Twenty such eggs, *fixed by drying,* were laid on a slide ruled in squares with a diamond and stained with aqueous Bismarck brown. Alternatively they may be fixed and stained in acetic-orcein, or fixed in acetic alcohol and stained in Feulgen.

Where the shells are thick (as in grasshoppers) they may be. removed by puncturing or by dipping for 2 min in 3% *aqueous sodium* hypochlorite. Otherwise the shells must be punctured if possible to admit the fixative and to aid the infiltration of wax.

Alcoholic fixatives, particularly Kahle's fluid or S.G. Smith's modification, are the best. Special methods of dehydration and infiltration are necessary for obtaining good sections, owing to the brittleness of the yolk. Smith's procedure is perhaps the most favourable. Haematoxylin or crystal violet (chromic method) staining are preferable for meiosis. Feulgen for cleavage divisions.

Saliary Glands

Salivary glands and some other bland tissues of Diptera contain nuclei in a perpetual interphase. These were first described by Balbiani in 1880 but only understood fifty years later. Their nuclei are assumed to be polyploid on three grounds: their size, the multiple or polytene structure of their chromosomes, and their total DNA content which is a high multiple of that of ordinary mitotic nuclei.

Their chromosomes, when fully extended, are about 100 times as long as at metaphase of mitosis. The 128, 256 or 512 threads of each homologous type are associated in parallel; each gene therefore appears as a band which is visible in the living state. The fully developed banded chromosomes are as long as a quarter or half a millimetre and when stained are visible with a hand lens.

Structural hybridity is shown by the same changes of partner as with pachytene pairing at meiosis. Salivary glands are important therefore both for study in combination with genetics and systematics and for micro-chemistry. Salivary glands can also be made use of in labelling experiments, such as with the use of radioactive precursors of DNA and RNA in demonstrating the metabolic activity of genes.

Larvae of the most suitable species may be obtained as follows

Chironomus is the common blood-worm of ponds.

Bibio feeds on roots and leaf-mould.

Drosophila the fruit fly, may be trapped by a bait of overripe fruit, and the flies bred on a special pabulum.

Sciara the dung fly, grows on fungi; can be bred on mouldy *Drosophila* pabulum.

Well-fed larvae ready for pupation have the largest chromosomes. Each salivary gland contains *28-32* cells in *Sclara,* 28-44 in *Chironomus, 100-120* in *Drosophila.* These are in different stages of development-the largest in the middle or at the rear end.

Dissection

Take the full-grown larva, having established its sex if necessary, and place it on a slide in a large drop of isotonic Ringer solution or stain fixative. Cut off its head with a needle in the right hand while pressing the body with a needle in the left. When the pressure is released the salivary glands will float out and can then be put on another slide. Separate preparations can be made of ganglia, guts, or gonads as required.

Tadpole Tails

The epidermal cells of the tail in the urodele larva (e.g., *Triturus,* the newt) are marked out for the study of mitosis. The cells are particularly favourable for prophase, since they are large and very flat. The best stage is when the tail fin is fully grown. Under natural conditions this stage is reached in May, June and July.

In the laboratory the larvae can be partially starved and brought into condition when they are needed. The tail should be cut off just behind the anus and fixed in acetic alcohol for two to twenty-four hours. Preparations can later be made in iron aceto-carmine or acetic-orcein, or with Feulgen.

Care should be exercised when using Feulgen not to press the preparation too hard lest the cells be crushed. Larvae with amputated tails can be retained and used once more when the tail has regenerated. In Great Britain a vivisection licence is required for such operations.

Amphibian Testes

To study meiosis in the newt, testes should be fixed about a month after the breeding season. They should be cut into small pieces and fixed in acetic alcohol for two or three days. They are best stained with iron aceto-carmine.

Very small pieces, the size being accommodated to that of the cover slip, should be placed on a slide in a small drop of stain, tapped out lightly, and covered after the material has darkened with the stain. The cover slip should be pressed lightly in the middle under blotting paper, and suitable isolated cells will be found lying at the edges of the preparation.

Lampbrush Chromosomes

Chromosomes of a type known as 'lampbrush' appear at a phase corresponding to diplotene in the ovarian oocyte nuclei of a wide variety of animals. Newts have the longest lampbrush chromosomes of all organisms and since these chromosomes are extended about twenty-five times the length of the chromosomes at metaphase of mitosis the longest lampbrush chromosome in *Triturus cristatus cam Vex* is about 700-800 μm long.

Lateral projections, consisting of pairs of loops some 20 μm or more when fully extended, are formed at chromomeres along the axis of these chromosomes. The growth and facility of loops to collect or discharge materials during growth -of the oocyte reflects a metabolic activity of genes at the loci at which they form. The organisation and fine structure of these chromosomes has been studied extensively. Owing to their enormous length and fragile nature lampbrush chromosomes are best observed under phase contrast after isolation in saline.

The germinal vesicle or nucleus from an ovarian egg is removed on to 0.1 molar KCl using forceps, needle and a binocular dissecting microscope. The nuclear membrane is freed from adhering yolk and cytoplasm by pipetting; the nucleus is then transferred to saline in a special observation chamber.

The observation chamber consists of a 3 in by 1 in slide with a ¼ in hole bored through the centre; a cover slip is placed across this hole and sealed to the slide with paraffin wax. For the newt *Triturus cristatus carnifex* a suitable saline is 7 parts 0.1 molar KCl: 3 parts 0.001 molar KH_2PO_4. In the observation chamber the nuclear membrane is removed using a pair of the finest watchmaker's forceps and a tungsten needle pointed in molten $NaNO_2$. The nuclear sap,disperses, the chromosomes become disentangled from one another and fall on the cover slip forming the bottom of the chamber.

The preparation is covered with a second cover slip sealed in position with vaseline, to prevent evaporation of the saline, and examined with a phase contrast microscope having its optical train inverted. Since the chromosomes are in. perpetual Brownian movement, photographs at high magnification must be taken with a flash.

Culex Testes

The pupa is placed on a slide in Ringer solution A and' examined by dissecting microscope. Needles are placed in the head and tail, and the larva is broken in two as the needles are pulled apart. The testes with their small ducts attached, if the individual is male, are readily visible as small translucent egg-shaped bodies.

They are removed with a fine needle to acetic alcohol, where they fix in two minutes. They should now be picked up by the needle and placed on a slide in a drop of iron aceto-carmine or acetic-orcein. The preparations can be covered after about one minute's staining (the time should be adjusted to suit stain and material) and pressed lightly through -blotting paper applied above the cover slip.

Preparations can be made permanent. This method is suitable for most small insect testes. An account of insect morphology may help in dissection,, e.g. Imms.

Blood

Red bone marrow is a useful source of mitosis in mammals.

Moreover its study can provide some understanding of the causal relationship between mitosis and blood diseases. Active marrow is best obtained from the young of birds and small mammals by extraction from femoral bones.

In man it is now obtained by sternal puncture. New improved techniques are now available for handling these cells. In small mammals colchicine can be injected intraperitoneally to accumulate metaphases and spread the chromosomes. This can then be followed by treatment of cells in hypotonic fluids which further aids spreading by expanding the cells. Excellent preparations can be obtained when this sequence of treatments is followed by fixation of the cells as a suspension.

A simple method involving direct colchicine treatment of aspirated marrow is described by Tjio and Whang. Peripheral blood is now the best source of mitosis in humans. It has been recently discovered that extracts from the seeds of *Phaseolus vulgaris* produce active cell multiplication after three days of incubation *in vitro.*

The active agent is phytohemagglutinin. Its mode of action is not yet properly understood. The blood cells which undergo mitosis are probably the monocytes and the lymphocytes. Blood is obtained by venipuncture and excellent preparations of mitotic chromosomes can be obtained by processing it according to Schedule 24.

Mammalian Embryos

Methods for obtaining chromosome preparations from mammalian embryos have been described by Ford and Woollam, Tarkowski, Shaver and Carr, Butcher and Fugo, Wroblewska and Dyban and Evans *et al.*. The last authors devised a rapid method for mouse embryos which utilises the membranes from implants of a gestational age of seven and a half or more days.

The implants are dissected from the gravid uterus and treated with colcemid before the membranes are removed for subsequent hypotonic pre-treatment. After brief fixation in acetic alcohol, the membranes are transferred to 60% acetic acid, where the cells become detached.

The free cells are then spread by drying the acetic acid on heated microscope slides. The spread cells are stained in toluidine blue-resin, a combined stain and mounting medium. The method when modified is applicable to tail tips of 1-6-day-old mice.

Tumours

Modern techniques provide rapid and satisfactory means of -studying mitotic activity in tumours for cancer diagnosis and therapy in man, mammals and birds. The simplest method is by acetic-lacmoid or orcein squashes. More elaborate methods are necessary, however, for detailed chromosome study. Siliconed slides can be used in order to facilitate chromosome spreading in cells of ascites tumours.

Better results are obtained if the cells are pre-treated in hypotonic fluids. The method suggested by Ford and Hamerton is suitable for all types of tumour cells where Feulgen staining is required.

Diagnosis of Sex in Somatic Tissue

The cytological diagnosis of the sex chromosomes in resting somatic nuclei has made possible a new approach to the study of normal and pathological human sexual development. In animals and insects the sex chromosomes,when large, form distinct chromocentres which are often located at the nuclear membrane.

Moore *et al.* found in biopsy specimens of human skin that female skin shows about 70% of nuclei with a single chromocentre (the X) and male skin about 5%. Many other human tissues have since been shown to present the same sex difference. Squash, smear or section methods, according to practicability, may be used to prepare the tissues for study. Acetic alcohol (1 : 3), 95% alcohol or methyl alcohol are suitable fixatives for squashes or smears.

A new fixative consisting of 30 parts 95% alcohol, 20 parts formalin, 10 parts glacial acetic acid, 30 parts distilled water is suitable for tissues prepared as sections. To avoid errors in diagnosis it is necessary to readily distinguish chromocentres from small nucleoli. Methods employing the Feulgen stain obviate such errors.

Alternatively, the thionin method employed by Klinger and Ludwig may be used. Pearson, Bobrow and Vosa have demonstrated that male nuclei can be positively identified at interphase by virtue of the fluorescence given by the Y chromosome when nuclei are stained with quinacrine dihydrochloride.

In nuclei of buccal smears, single fluorescent bodies were observed in normal XY males and double fluorescent bodies in XYY males. However, such identification is limited to man and gorilla,, since it is only in interphase nuclei of these species that brilliant quinacrine fluorescence of the Y chromosome is obtained.

The quinacrine-fluorescent spot in somatic cells of human males can also be seen in 50% of the sperm from normal individuals and has been claimed to represent the Y chromosome. Some observations concerning the findings in sperm are discussed by Evans.

Giemsa Banding in Metaphase Chromosomes

A Giemsa staining method that preferentially stains centromeric heterochromatin in mouse chromosomes was first described by Pardue and Gall. Its specificity is believed to depend on denaturation of the DNA in fixed, air-dried preparations followed by annealing in warm saline prior to staining.

The centromeric DNA in mouse is known to consist of highly repetitive base sequences and to anneal more rapidly than the non-repetitive DNA of the genome. This so-called C-banding technique has been variously modified to demonstrate both centromeric heterochromatin and other characteristic Giemsa (G-) banding in human and other mammalian chromosomes.

It can also be used on meiotic chromosomes. The factors and mechanism involved in the induction of specific banding patterns in mammalian chromosomes have been examined by Kato and Moriwaki, Comings, Avelino, Okado and Wyandt, Sumner, Evans and Buckland and Sumner and Evans.

Although it is clear that G-banding does not have to be associated with DNA made up of repetitive base sequences the Giemsa stain is seemingly bound exclusively to DNA. The reason for the disparate behaviour along the chromosomes remains obscure. An exact correspondence in G-banding pattern has been produced in human chromosomes by the use of trypsin and other proteolytic enzymes prior to Giemsa or Leishmann staining.

Similarly, specific banding is inducible in human chromosomes by means of potassium permanganate prior to Giemsa staining.

Finally, Dutrillaux has shown that preferential staining of the characteristic bands can be reversed (R-banding) in mammalian chromosomes by treatment of the chromosomes in spread preparations with Earles' medium at pH 6.5 for 1C-20 minutes at 87°C prior to Giemsa staining.

In other words, with R-banding, it is the remaining bulk of the chromosomes which becomes intensely stained. G-banding was at first more difficult to apply to plant chromosomes inasmuch as conventional N HCI hydrolysis, so valuable in obtaining flattened meristematic cells in squash preparations, inhibits selective Giemsa staining.

For the same reason cell separation in pectinase or snail cytase is unsuitable. Good results have been obtained, however, in squash preparations of root tips by using 40 or 45% acetic acid for maceration or following hydrolysis for 25-30 seconds in N HCl at 60°C. More consistent results are obtained when fixed material is steeped in 45% acetic acid for 15-30 minutes at 60°C prior to maceration.

Excellent preferential staining of bands is obtained in squash preparations of plant chromosomes when the DNA is denatured in a saturated solution of barium hydroxide for 5-8 minutes at room temperature prior to annealing in warm saline.

Schweizer found an exact correspondence in the position of Giemsa bands and H-segments previously noted in the chromosomes of certain plant species *e.g. Trillium grandiflorum* and *Fritillaria lanceolata*. Giemsa banding is evident also at prophase in both plants and animals, and chromocentres are preferentially stained at telophase and interphase.

Recognition of Heterochromatin with Fluorochromes

The use of fluorochromes in the study of the linear organisation in chromosomes was first described by Caspersson, Foley *et al.* and Caspersson, Zech *et al.* They observed that intense fluorescence occurred regularly at specific loci in the chromosomes of certain plants, when preparations were stained with acridine derivatives and especially quinacrine and its mustard.

The position of these loci in *Trillium* chromosomes was found to more or less correspond with those of H-segments described by Darlington and La Cour. Vosa has since studied the fluorescence given by various fluorochromes in the heterochromatin of several plants. He classified four classes of heterochromatin, according to whether they gave '*enhanced*' or '*reduced*' fluorescence with quinacrine and responded negatively or positively to cold treatment.

The phenomenon of enhanced and reduced fluorescence in H-segments is as yet not fully understood. The investigations have suggested that the fluorescence of quinacrine, in the presence of DNA, is intensified where there is relatively high AT content and quenched where the DNA is correspondingly rich in GC.

The two categories of fluorescence may, however, reflect differences in states of chromatin condensation peculiar to two main classes of heterochromatin. There is clearly, though, a close correspondence between Giemsa banding and the segments distinguishable by quinacrine fluorescence, irrespective of category.

The fluorochromes, as 0.5-1 % aqueous solutions, can be applied to squash preparations of testes, salivary glands and neural ganglia of Diptera, suspensions of mammalian cells prepared as films, as well as squash preparations of RT macerated and squashed in 40% acetic acid. The stained cells are examined in temporary preparations mounted in distilled water. A microscope fitted for fluorescence microscopy, with a mercury lamp, excitor and barrier filters is required.

In squash preparations, particularly of plant material, incident illumination is usually best because background debris is then invisible.

Identification of Sister Chromatids by Differential Staining

Methods have recently been developed which allow the :identification of the two cbromatids in chromosomes by differential staining and consequently the observation of sister chromatid exchanges. This development stems from the observations of Zakharov *et al.* showing that the sister chromatids of chromosomes in mammalian cells become disparately condensed at metaphase, after two rounds of replication in the presence of 5-bromodeoxyuridine.

The disparity, evidently a legacy of bifilar and unifilar substitution in the DNA, is *further* recognisable by the differential, staining behaviour that is obtained with Giemsa and the fluorochrome Hoechst. The Giemsa and Hoechst dye, respectively stain and fluoresce weakly in the under-condensed chromatid arising after bifilar substitution, and conversely so in, the fully condensed and unifilarly substituted sister. The result is a distinct harlequin effect.

The fluorescence produced by the Hoechst dye is unfortunately unstable and fading is rapid in light. Differential staining of the sister *chromatids* can, however, also be observed after staining with acridine orange, but the differences betweena chromatids is not readily visible until the cells are exposed to exciting light passed through a BG12 filter.

Differential fluorescence is not obtainable with the fluorochrome quinacrine. A technique perfected by Perry and Wolff, in which: the use of the Hoechst dye is combined with Giemsa staining,. to make permanent preparations is given in Schedule 27.

In Situ Localisation and Characterisation of Different Classes of Chromosomal DNA

It has been found that chromosomal regions containing satellite DNA stain distinctively from other parts of the chromosomes, after staining with a 0.5% solution of acridine orange in McIlvaine buffer (pH 7), subsequent to denaturation of the DNA with alkali.

The distribution of double-stranded and single-stranded DNA in the chromosomes *in situ* can be assessed from the colour of the fluorescence of acridine orange in light with a wavelength of approximately 510 nm. Double-stranded DNA fluoresces green and single-stranded DNA red.

Denaturation of the DNA with alkali in metaphase chromosomes results in a shift in colour of fluorescence from green to yellow, orange, brown and red, as denaturation progresses.

In chromosomes of mouse and the vole, *Microtus agrestis,* the DNA denatures uniformly with alkali and in chromosomes of human cells sequentially. Further, after prolonged reassociation in SSC saline, areas containing repetitious DNA, as in mouse centromeres, fluoresce more intensely with acridine orange than other parts of the chromosomes.

Embryo Sac and Endosperm

Meiosis in the embryo sac may take place as long as two or three weeks after meiosis in the another. In *Fritillaria* and *Tulipa* it usually coincides with the pollen grain division.

Wherever possible the ovules should be removed from the ovary before fixation. In a number of Monocotyledons they come out in entire strings. Very large ovaries should have part of the ovary wall removed.

The squash method after acetic alcohol fixation, either with Feulgen or acetic-lacmoid staining, is most suitable where cells and chromosomes are large. In favourable plants it is possible to tap out the cells, e.g., in 45% acetic acid after Feulgen staining, and remove unwanted tissue with the aid of a dissecting micsoscope before squashing. For small chromosomes and material difficult to handle the paraffin method has to be used, with a strong and penetrating aqueous fixative, e.g., 2BX.

Endosperm divisions are to be found a few days after pollination. In some species, e.g., *Scilla sibirica* and *Lilium* the division are synchronised for about six mitotic cycles and later with all formation partially synchronised in patches.

Endosperm tissues are best handled as Feulgen squashes, the tissue being dissected out entire after staining. Dissection is performed under a dissecting microscope, using finelypointed tungsten needles. The stage of mitosis and intensity of stain can be determined during this operation. Weak staining can be corrected by applying acetic-orcein to the dissected tissue. Longer hydrolysis times than those required for other tissue are necessary for some species and, in any case, after long storage.

In Coniferales it is possible to dissect entire endosperms from female cones before fixation. At the right stage they look like the pupae of ants and reach this stage in northern countries during May. Fix in ' acetic alcohol and stain as squashes with acetic-orcein.

Pre-Meiotic Mitosis

Pre-meiotic mitoses are to be found in the earliest stages of anther and embryo-sac development. In the anthers the mitoses are synchronised at least locally. Fix dissected organs, where possible, in acetic alcohol and treat either as Feulgen squashes or stain as squashes with acetic stains.

Pachytene

The pachytene stage is pollen mother cells has been used for exact comparison of chromomeres, centromeres, nucleolar organisers and heterochromatin in many flowering plants. Special techniques have been developed. The following are representative examples

Centromeres	*Agapanthus*	Darlington
Centromeres (diffuse)	*Luzula*	S.W. Brown
Chromomeres	*Secale*	Lima-de-Faria
Nucleolar Organiser	*Zea mays*	McClintock
Heterochromatin	*Cicer*	Thomas
Heterochromatin	*Lycopersicum* S.W.	Brown
Pairing : 2*x*	*Zea mays*	McClintock
Pairing : 4*x*	*Hyacinthus*	Darlington
-Pairing : 4*x*	*Tulipa*	Upcott
Pairing : 4*x*	*Lycopersicum*	Gottschalk

8

Staining Procedures and Histological Stains

Many compounds are coloured but not all of them are stains. A stain or dye is a coloured compound that can be bounded by a substrate. In histological staining dyes are used to impart colours to the various components of tissues. Some-times the colour procedure has a high degree of chemical specificity, so that a stain can be used as a histochemical reagent.

There are three general methods of staining the first of these is "*direct staining*", in which soaking the objects or sections in the stain is sufficient to colour the desired structure. This is rarely employed and can be used only with those stains which are entirely specific for the structure which is to be emphasized.

It is far more usual to employ a process of "*indirect staining*" in which the whole object or section is first uniformly stained in the dye selected and the "*differentiated*" in some solution which removes the dye from all those structures other than the once which are to be emphasized. The third method of stain-ing is "mordant staining" in which some preliminary solution is used which causes the stain to "*bite*" into a structure with-out being obsorbed by others.

These three great categories are by no means sharpely divided because many of the so-called direct stains actually incorporate mordants, and many of the indirect staining solutions may be used by direct method if they are first greatly diluted with water. Before going for details of staining, one must have the knowledge of the effects of fixative on stains.

EFFECTS OF FIXATIVE ON STAINS

It is not so much the chemicals of the fixatives themselves that affect the later staining of various materials as the chemi-cal nature of the various tissues plus the compounds resulting. from chemical reaction involving these tissues and the reagents in the killing and fixing fluids which, to a great extent deter-mine the results that are obtained with different dyes.

Let us take some examples:

(*a*) Safranin has a staining affinity for chromatin Fluid containing chromic acid and formation, perhaps by a sort of mordanting action, leave chromatin containing structures such as nuclei and chromosomes in a nearly perfect condition for staining with safranin. If, however, chromic acid reacts with substance in the cells to leave them coloured dark brown, the safranin is correspondingly dulled, the dullness and brownness increasing with the intensity of the reaction between the acid and the cell contents.

A point will be reached where there is no longer any differentiation between cytoplasm and other cell contents (as when the tissue are saturated with tannin compounds) and bleaching as a prelude to staining becomes necessary.

(*b*) Some killing fluids pressure the plastin of mitotic figures perfectly, of these fluids some mordant the plastin so that if stains but other fails to mordant it. This explains why certain staining procedures which are supposed to reveal the so-called "*Spindle fibres*" occasionally fail, the fault is not because the staining was not carried out perfectly but because the structures involved were inadequately mordanted by the preservatives. Navashin's fluid, for example, preserves plastin but rarely mordants it.

(*c*) The triple combinations work as they should only on material originally fixed in a fluid that contains osmic acid or on sections that were mordanted in a weak solutions of osmic acid in chromic acid. Otherwise, the chromosomes tend to appear purple rather than red, i.e., they have an affinity for violet rather than for safranin.

Bleaching

Sections cut from materials which were exclusively darkened by the fixing fluid require to be bleached before a proper stain can be obtained. One should not, however, bleach sections without being fairly certain that it should be done because bleaching frequently is more or less damaging the tissues. Many types of materials which are naturally dark (such as the mechanical tissue in fern rhizome), do not actually require bleaching and can be beautifully stained if a proper selection of dye is made

(*a*) If the section appear to have suffered from too much chromate fixation, immerse the section far one minute in a 1% aqueous solution of potassium permanganate, rinse, remove the permanganate in a I % aqueous solution of oxalic acid, wash again and proceed to the staining.

(*b*) Chlorine gas fumes bleach excellently. Place enough crystals of potassium chlorate to approximate the size of a grain of wheat in coplin and pour a little diluted hydrochloric acid. As soon as the greenish-, ellow fumes starts coming, fill the jar quickly with 50% alcohol. About 20 minute minimum immersion of the sections is required to effect bleaching. Sections, however, cannot be stained if commercial bleaching fluids containing chlorine are used.

(*c*) Boric acid solution has been used for bleaching as has a 2% aqueous solution of ammonium persulphate. The latter should be used with caution as it is a powerful oxidiser.

To restore the staining capacity of tissues bleach if neces-sary, then soak for fifteen minutes in a 10% solution of benzol peroxide in acetate, wash out in a solution of 2 parts xylol to 3 parts of accetate followed by absolute alcohol.

Differential Acidification

In order to identify and localise stains in tissues whose elements do not present sufficient contrast in their relative degree of acidic or basic reaction, the hydrolysis phase of Fenlgen's reaction may be employed to render nuclei' more acid and thus to give them a greater affinity for nuclear stains.

The procedure is especially adaptable to more or less thick materials which are to be mounted entire-such as fern, prothallia, leaves and similar other objects. The slide or fixed materials are brought down to water, placed in cold 1/N HCl for exactly one hour, rinsed with one change of distilled water and then placed in the nuclear stain. The acid must not be heated as is done in the regular Feulgen technique, otherwise the material is very likely to become dissociated.

Progressive and Regressive Staining

By observing the process of the staining under the micro-scope from time to time, any desired intensity may be attained. This is known as "*progressive staining*". A sharp differentiation usually cannot be obtained by this method. Consequently, the general practice is to overstain considerably and then to destain or otherwise to differentiate until a satisfactory optimum has been reached. This is known as "*regressive staining*". Practically, all the schedules given in the present chapter concern regressive staining procedures.

General Staining

A general stain is one that stains everything indiscrimina-tely. There is little or no selective differentiation, but such may to a slight extent be obtained by regressive destaining.

Classification of Dyes used in Microscopy

With few exceptions, notably haematoxylin and carmine, the stains used for colouring microscopic tissue preparations are synthetic dyes. These dyes constitute a broad class of aromatic organic chemicals which can be divided and sub-divided into various groups according to their chemical and physical properties. Brief descriptions of some of these divisions are given below

Acid Dyes

Strictly speaking, dyes of this class are those in which the balance of the charge on the dye-ion is negative. Acid dyes, with some exceptions, stain tissue elements that are basic in reaction.

Basic Dyes

A dye of this class is one in which the charge on the dye-ion is positive, but not all so-called dyes are ionic. Basic dyes stain tissue elements that are acidic in reaction, but there are also exceptions.

Amphoteric Dyes

Under certain conditions a number of dyes are amphoteric. This occurs among both acid and basic dyes. Many acid dyes possess both reactive acidic groups and reactive basic groups, in addition to their cations. These groups together confer amphoteric properties on the particular dye.

Such dyes are normally regarded as being acid dyes. With one exception, there are no basic dyes whose molecules possess both reactive basic and reactive acidic groups, in addition to their anions, the exception being nono-fuchsinic acid, recently discovered. This somewhat bizarre compound might be regarded as a strongly acid basic dye.

Neutral Dyes

Most of these are not really dyes at all, but are non-ionic organic colouring matters. Some of them contain acidic or basic groups, however, and for this reason such neutral "*dyes*" are sometimes classified as "*acid*" or "*basic*". Among the neutral "*dyes*" are the Sudan colours (1, 2, 3, 4 etc.). This type of "dye" is probably absorbed or dissolved by lipid to which colour is imparted. A few non-ionic "dyes" such as Sudan Black, which has two basic groups (inino), might con-ceivably unite chemically with acidic elements of certain tissues.

All "*neutral dyes*" can be converted into true dyes by the process of sulphonation. They then become water-soluble dyes. In the strict sense a dye is a coloured aromatic, organic substance capable of electrolytic dissociation which enables it to unite by chemical and/or physical means with tissue-elements or other substances of opposite charge.

Neutral dyes should not be confused with "neutral" or compound stains. The latter are formed by chemical union between a basic and an acidic dye. Such compounds of basic and acid dyes have long been known to histrologists. It is doubtful, however, if the average textile dyestuff manufacturer is aware of such preparations. It is still more doubtful if they would be of any interest to textile dyer.

Compound Dyes

As stated above these are neutral dyes of the kind that appear to be known only to the histologist. They are produced by neutralising suitable acid dyes with suitable basic dyes. That is to say, the negatively charged dye-ion of the acid dye units with the positively charged dye-ion of the basic dye when aqueous solutions of the two dyes are brought together, to form a precipitate which is a two-colour dye.

If, however, one of the dyes is a polychronic stain, then the resultant compound stain will also possess the same polychrome properties plus the colour of the other dye. Such a compound dye is a true chemical compound and not a mixture. Among such dyes are Jenner stain *Gunner,* Leishman stain *Leishman,* methyl green organge *G. Kardos,* neutral red-light green *Twort* and neurtal red-jast green *FCF Ollet*. Another series of compound dyes known to histologists consists of a mixtures of acid dyes of the Masson types.

But these are not chemical compounds. Then there are mixtures of basic dyes such as methyl green-Pyronin. These are mixtures, not chemical compound. Although it is possible to form chemical compound dyes by the union of many basic dyes with many acid dyes; as well as by the interaction of certain acid dyes with certain other acid dyes, it is not possible to produce chemical compound dyes by the interaction of any basic dyes with any other basic dyes.

The reason for this is that there are no basic dyes, as there are acid dyes, whose molecules contain both reactive basic and acidic groups on the dye-ion itself.

Mordant Dyes

By colour index (convention) the mordant dyes are defined as those used in conjunction with metal salts. The mordant maybe applied before the dye, together with the dye as a soluble *dye-metal* complex or after the dye. In some instances a mordant dye is also an anionic or, rarely, a cationic dye in its own right. *Mordant dyes* have many uses in histology and histochemistry.

Some of the important and frequently used stains are given below:

Method 8.1

Haematoxylin and Eosin

The method described here is typical of many H and E procedures.

Solution required

A. Mayer's haemalum

Dissolve the following, in the order given, in 750 ml of water.

Aluminium potassium sulphate	-50 gm
Haematoxylin (C.I. 75290)	-1.0 gm
Sodium iodate	-0.1 gm
Citric acid (monohydrate)	-1.0 gm
Chloral hydrate	-50 gm

Make up with water to 1000 ml. The solution may be reused many times. If the solution fails to stain nuclei pro-perty, it may be over-oxidised.

B. Eosin

Eosin (C.I. 45380)	-2.5 gm
Water	-500 ml

This solution keeps indefinitely and may be used repeatedly. Moulds often grow in it and need to be removed by filtration. Alternately, use 0.2% ethyl eosin (C.I. 45286) in 95% alcohol.

Procedure

1. De-wax and hydrate paraffin sections. Frozen sections should be dried onto slides.
2. Stain in Mayer's haemalum (solution A) for 1-15 minutes (usually 2-5 mts). Overstained sections can easily be differentiated by agitating for a few seconds in 1% (v/v) concentrated hydrochloric acid in 95% alcohol, then washing throughly in tap water.

3. Wash in running tap water for 2 or 8 mts or until the sections turn blue. If the tap water is not sufficiently alkaline to blue the sections, add a few drops of ammonium hydroxide or of saturated aqueous lithium carbohydrate or a small pinch of calcium hydroxide to about 500 ml or water and leave the washed sections in this for 3C-60 sec, then rinse in tap water again. Examine the wet slide under a microscope to check that selective nuclear staining has been achieved. Any blue colouration of cytoplasm and connective tissue should be extremely paint.
4. Immerse slides in eosin soln for 30 sec. with agitation.
5. Wash and differentiate in running tap water for about 30 sec.
6. Dehydrate in 70%, 95% and two changes of absolute ethanol (with agitation, about 30 sec. in each change; without agitation, 2-3 mts in each change).
7. Clear in xylene and cover, using a resinous medium.

Result

Nuclear chromation-blue or purple, cytoplasm, collagen, keratin, eryturocytes-pink.

Method 8.2

Ehrlich's Haematoxylin

Solution required

Haematoxylin (Ehrlich)

Potassium permanganate - 0.1

Procedure

1. Material should be fixed in 10% formalin and embedded in paraffin wax.
2. Fix sections to slides and bring down to distilled water as usual.
3. Stain in Ehrlich haematoxylin for 10 mts.
4. Pour off excess stain, then rinse and blue in tap water.
5. Immerse in potassium permanganate 0.1 % for 10 sec., then wash well with water.
6. Dehydrate; clear; mount in balsam.

Results

Keratohyalin is stained blue-back while the other elements, are unstained or faintly stained.

Method 8.3

Mallory's Phosphotungstic Acid Haematoxylin

Solution required

Haematoxylin 10% in absolute alcohol (ripened for 3 months or more)	-1 ml
Phosphotungstic acid	-2 gm
Distilled water	-100 ml

Procedure

1. Fix in Zenker Embed in paraffin wax.
2. Bring sections down to distilled water.
3. Treat with iodine to remove mercuric precipitate.
4. Remove iodine with 0.5% aqueous sodium hyposulphite.
5. Wash thoroughly in running water:
6. Immerse for five to ten minutes in 0.25% potassium permanganate; then wash in tap water.
7. Immerse for ten to twenty minutes in 5% oxalic acid; then wash thoroughly with tap water.
8. Stain twelve to twenty-four hours in haematoxylin solution prepared as above.
9. Wash in tap water; dehydrate with 95%, and absolute alcohol.
10. Clear in xylol and mount.

Results

Nuclei, centrioles, achromatic spindles, fibroglia, myoglia, fibrin, contractile elements of straited muscle-turn blue. Collagen, reticulum, ground substances of cartilage and bone-turn yellowish to brownish red. Coarse elastic fibrils turn faint purple.

Method 8.4

Harris' Haematoxylin

Solution required

Haematoxylin crystals	-2 gm
Aluminium chloride	-1 gm
Alcohol 50%	-1000 ml

Heat on a waterbath until dissolved. Add 6 gm mercuric oxide and fitter. Add to the filtrate 1 ml conc. HCl.

Procedure

1. Bring slides down to water, or wash the killing fluid out of the material with water.

2. Rinse for about 20 minutes in the haematoxylin solution.
3. Rinse in distilled water until all the excess stain has been washed away.
4. Destain section in acid water (about 5 drops or less of HCl to each 100 cc of water). As a rule the time is about 5 sec., but one can easily stop the action of the acid, rinse the material with tap water (which will blue the stain).
5. Wash for a few seconds in tap water. If water is not sufficiently alkaline to blue the materials, and 1 or 2 drops of ammonia to the water. If it is desired to clear the cytoplasm, place the slide or materials in a weak solution of lithium carbo-nate in water. About 10 mts suffices.
6. Rinse in water again, then proceed with the dehydration and mount as usual.

Result

Useful stain for plant materials such a fern prothallus, the Rhodophyta, many fungi and similar objects.

Method 8.5

Delafleld's Haematoxylin

Solution required

Haematoxylin crystals	-1 gm
Absolute alcohol	-6 ml
Ammonia alum	-100 ml
Glycerine	-2.5 ml
Methyle alcohol	-2.5 ml

Procedure

1. Dissolve haematoxylin crystals in absolute. Add this drop by drop to 100 ml of saturated solution of ammonia alum. Expose to light in open bottle for one week. Filter and add glycerine and methyl alcohol. Pour into a shallow dish and expose to a quartz mercury vapour lamp for 2 hrs.
2. Transfer the material to the stain fern either water or 50% alcohol. The length of time for which the sections may be left in stain depends parsy on the character of the material, partly on the nature of the killing fluid. Roughly 3-30 mts is time taken.
3. Wash in running tap water a few minutes to remove completely

all excess stains. The washing should be thorough to avoid the precipitation.

4. Treat with aciduate water, and then transfer quickly to water and wash in running tap water until the section acquire a rich purple colour.
5. Pass through 50%, 70% and 95% alcohol.
6. Rinse in xylol and mount in balsam.

Method 8.6

Heidenhain Iron Haematoxylin

Solution required

(A) Iron alum	–4 gm
Distilled water	–100 ml
Dissolve by shaking. Do not heat.	
(B) Haematoxylin 10% in absolute alcohol (ripened for 3-6 months)	–5 ml
Distilled water	–95 ml

Procedure

1. Mordant sections in A for to 24 hrs, the time varies with material and fixative.
2. Rinse with water.
3. Transfer to solution B and allow to remain the same time as in stage 1. The sections should not be an uniform pitch black and show no details.
4. Rinse in water.
5. Differentiate in solution A; control destaining under the microscope.
6. Wash in running water for 5 mts
7. Counter stain if desired.
8. Dehydrate, clear and mount.

Remaining stains are mentioned in details in different chapters to follow.

9

SPECIFIC STAINING FOR CELL ORGANELLES

The study of cells has assumed much significance for biology students as well as research scholars and it is no longer enough just reading about it. Students specialising in cell biology are actually expected to follow various sophisticated pchniques. Some of the simple techniques described in the following pages are concerned with the cell organelles and these are essential in staining them for both qualitative and quantitative recordings.

STUDY OF MITOCHONDRIA

Any tissue can be selected for the study of mitochondria. The tissue is generally fixed in neutral formalin for 24 hours at room temperature followed by post-chroming for 3-24 hours at 37°C or in Legaud's fixative (4 vols. of 3% aqueous potassium dichromate and 1 vol. of 4% formaldehyde, mixed just before use) for 4-24 hours at 8°C followed by post-chroming as in neutral formal and washing overnight in running tap water.

Method 9.1

Fuchsin Acid-Light Green Method

(*a rapid method for staining mitochondria*)

Solutions required

A. Hydrochloric acid N/1

B. Fuchsin acid, 1% aqueous

C. Light green SF, 1% aqueous

Procedure

1. Paraffin sections of formalin-fixed material are mounted on slides and dewaxed with xylol as usual.
2. Pass the slides through the usual graded alcohols into water.
3. Hydrolyse for three minutes in the hydrochloric acid solution A), at 60°C.
4. Rinse in water.
5. Stain in the fuchsin acid solution for thirty seconds. 6. Rinse in water.
7. Counterstain in the light green solution for one to three minutes.
8. Rinse in water.
9. Dehydrate with 95% and absolute alcohol. 10. Clear in xylol.
11. Mount in D.P.X. or in Canada balsam in xylol.

Results

The mitochondria stand out sharply stained purplish red with green peripheral wall. Chromatin and collagen are green. Muscle tissue : purplish. Erythrocytes : brilliant red.

Note

It may be interest to compare the results obtained with this technique with those obtained using the same dyes, but under simpler conditions and without hydrolysis, in the techniques of *MacConaill* and *Gurr.*

Method 9.2

Fuchsin Acid (Altmann)-Methyl Green Method

(for mitochondria in Leucocytes)

Solution required

A.	Potassium dichromate	15 gm
	Distilled water	100 ml

B. Iodine 0.5%in 70% alcohol

C. Sodium thiosulphate 0.5% aqueous

D. Altmann's acid fuchsin

E. Methyl green, 1 % aqueous

Procedure

1. Make smears of fresh, unoxalated blood on scrupulously clean slides, and fix by placing them at once in Helly's fluid and leaving therein for twenty-four hours.

2. Post-chrome in solution A for forty-eight hours.
3. Immerse in solution B for two minutes.
4. Rinse in 70% alcohol.
5. Immerse in the sodium thiosulphate solution for one minute.
6. Wash in running water for several hours or overnight.
7. Transfer to distilled water.
8. Take a slide from the distilled water and blot or wipe away the water from all parts of the slide except that covered by the smear.
9. Place the slide over the corner of a tripod, flood with solution D and beat the preparation over a flame until fumes appear.
10. Pour off excess stain and wash with water.
11. Stain for a few seconds with methyl green by allowing a few drops of the solution to flow over the slide.
12. Dehydrate in two changes of absolute alcohol.
13. Clear in xylol and mount in Cristalite or D.P.X.

Results

Mitochondria, erythrocytes, and the specific granules of eosinophilic leucocytes are stained red. Nuclei are stained by the methyl green, while the cytoplasm also takes this dye but to a lesser extent.

Notes

(a) The use of oxalated blood is not to be recommended due to the alteration produced, by the commonly used oxalates, in the cytoplasm and nuclear structure of the leucocytes.

(b) Fixation of the blood smears before they are dry is absolutely essential, according to the author, to obtain well differentiated preparations.

Method 9.3

Fuchsin Acid—Picric Acid Method

(After Altmann)

Solutions required

A. Altmann's Fluid

Potassium dichromate 5% aqueous	1 volume
Osmic acid 2% aqueous	1 volume

Note: Although the penetration power of this fixative is poor, its very satisfactory for surface fixation.

B. Acid fuchsin 20 gm
Aniline water 95 ml

C. Picric acid saturated in absolute alcohol.

Procedure

1. Small pieces of tissue, not more than 2 mm. in diameter, are fixed for twenty-four hours in Altmann's Fluid.
2. Wash for an hour in running water; then dehydrate; clear, and embed in paraffin wax in the usual manner.
3. Sections, not thicker than 4 s, are brought down to distilled water; then stained for six minutes in Solution B.
4. Pour off excess stain; then blot section carefully; then differentiate and counterstain by flooding the preparation with Solution C.
5. Rinse quickly in 95% alcohol; then dehydrate with absolute alcohol; clear in xylol, and mount.

Results

Mitochondria are stained crimson against a vivid yellow background.

Method 9.4

Fuchsin Acid -Toluidine Blue-Aurantia Method

Solutions required

A. Champy's Fluid
Potassium dichromate 3% 7 ml
Chromic acid 1 % 7 ml
Osmic acid 2 % 4 ml

B. Pyroligneous acid 1 volume
Chromic acid 1 % 2 volumes

C. Potassium dichromate 3%.

D. Acid fuchsin 10 gm
Aniline water 100 ml

E. Toluidine blue 0.5% aqueous.

F. Aurantia 0.5% in 70% alcohol.

Procedure

1. Tissues are fixed in Champy's fluid for 24 hours; then washed in running water for at least one hour.
2. Immerse in solution B for 12-24 hours, then wash in distilled water for 30 minutes.

3. Mordant in solution C for three days : then wash in running water for 24 hours.
4. Dehydrate; clear; embed in paraffin wax.
5. Sections, 3-5 *a* in thickness are taken down to water in usual manner; then stained for 6 minutes by flooding the slide with solution D and heating gently till vapour rises.
6. Rinse iii distelled water; then counterstain solution E.
7. Rinse with distilled water; then stain with solution F for 30-50 seconds.
8. Differentiate with 95% alcohol; dehydrate; clear in xylol and mount.

Results

Mitochondria are stained red; nuclei-blue; background yellow.

Method 9.5

Phospho-Tungustic Acid Hematoxolin (PTAH) Method (Mallary)

Solutions required

1. Lugol's iodine solution
2. Aqueous solution of sodium thiosulphate 0.5%.
3. $KMnO_4$ aq. solution-0.25%
4. Aq. Oxalic acid solution-5%
5. Phospho-tungustic acid Hematoxylin

Hematoxylin	1 gm
Phospho-tungustic acid	20 gm
Distilled water	1000 ml

Dissolve the solids in separate portions of water, the hema-toxylin with the aid of gentle heat. When cooled, combine. No preservative is necessary, spontaneous ripening requires several weeks. It can be accomplished at once by adding 0.177 gm of $KMnO_4$.

Procedure

1. Paraffin sections of Helly fixed material are brought to water through toluene and alcohols and placed in a half strength Lugol's iodine solution until yellowish brown. Wash off excess iodine in water and dip in and out of 0.5% aqueous solution of sodium thiosulphate until the section is colourless. Wash several minutes in running tap water.
2. Place sections in a freshly made up 0.25% aqueous solu-tion of $KMnO_4$ for 5 minutes.

3. Wash in water.
4. Put in oxalic acid 5% aqueous solution for I ml or until clear and colourless.
5. Wash in running tap water for several minutes.
6. Stain in phosphotungustic acid hematoxylin for 12-24 hrs. Overnight staining (15 hrs) is found quite adequate.
7. Blot sections dry with paper towel or filter paper and pass rapidly through 95% and absolute alcohol, agitating the slides and shortening the time as much as possible since alcohol will extract the red part of the stain.
8. Clear in xylol and mount.

Results

Mitochondria of liver and kidney cells stained deep blue.

Method 9.6

PAS-PTAH Method

Solutions required

1. Regaud's Fixative;

3% aqueous Pot. dichromate	80 nil
Formalin	20 ml

2. 3% Pot. dichromate solution.
3. Altmann's Aniline Acid Fuchsin

Aniline	1 ml
Acid Fuchsin	4 gm
Distilled water	20 ml

Shake aniline with distilled water for a couple of minutes Le stand and then shake again several times at intervals over 24 hrs. Filter. Add 4 gm of acid fuchsin and shake at intervals over several hours. The solution keeps only long enough to be used the same day it is made.

4. Dilute sodium carbonate 0.1 % aqueous.
5. 1 % Hydrochloric acid.
6. 0.5% aqueous methyl blue.

Procedure

1. Fix tissue in Regand's fixative for four days, making up new Regaud's each day. Post chrome for 8 days in 3% Pot. dichromate changing every day.
2. Wash in running water for 24 p.

3. Dehydrate in graded alcohol, clear in toluene and mount in paraffin. Cut paraffin sectio, 2.4 μ thick.
4. Bring paraffin sections to water and place in steaming Altmann's aniline acid fuchsia for 5-10 minutes removing the beat from the solution, when the slides are placed in it.
5. Differentiate in dilate sodium carbonate until cytoplasm is pale pink and nearly colourless.
6. Stop differentiation and heighten colour by dipping briefly in 1 % HCl.
7. Wash in distilled watei and counterstain in 0.5% aqueous methylene blue. This will take only a few seconds.
8. Rinse in distilled water, dip in 1 % HCl briefly.
9. Wash in distilled water, dehydrate in alcohol.
10. Clear in toluene and mount in balsam.

Results

The mitochondria become a bright red nuclei become blue to green coloured.

Method 9.7

Van Gieson's Stain Method

Solution required

1. Van Gieson stain

Acid Fuchsin 1 % aq.	=10 c.c.
Picric acid saturated aq. solution (about 1.22%)	=100 c.c.

2. Alcoholic Hematoxylin

Hematoxylin	= 0.5 gm
95% alcohol	=10 c.c.
Distilled water	=90 c.c.

Allow this stain to ripen for 4-5 weeks then dilute this with equal parts of distilled water.

Procedure

1. Bring sections to 95% alcohol.
2. Place in alcoholic iodine for 5 mts. to remove mercuric chloride.
3. Rinse in 80% alcohol or wash in tap water.
4. Transfer to 5% sodium thiosulphate for 2 minutes to bleach iodine.

5. Wash well in running tap water.
6. Mordant in ammonia ferric alum (2.5%) solution for 1-2 hrs.
7. Wash quickly in distilled water.
8. Stain for 1-36 hrs in alcoholic hematoxylin.

Note. The section should be adequately stained when on being taken out of the hematoxylin, they are a homogenous jet black and microscopically show no cellular detail at all.

9. Wash in water.
10. Differentiate in 2.5% Iron alum. The sections should be rinsed in tap water before each examination, which will immediately stop the decolourisation.
11. Wash in running water 15-60 mts.
12. If desired counterstain with Van Gieson's stain or eosin.
13. Rinse with 50% alcohol.
14. Dehydrate in two changes of 95% alcohol.
15. Two changes in 100%, three in xylol and then mount.

Results

Mitochondria, chromatin, centrioles, nuclei are stained black.

STUDY OF GOLGI COMPLEX

The Golgi complex is studied in sections by using different techniques. The tissues which can be easily handled for Golgi preparation are epidiclymus, small intestive, brain, spinal cord, pancreas. Some of the staining methods for Golgi complex are given below:

Method 9.8

Silver-Nitrate-Gold Chloride—Paracarmine Method

Solution required

A. Cobalt nitrate 1 % aqueous 100 ml.
   Formalin 15 ml.

B. Silver nitrate 1.5% aqueous.

(N.B. : This should be stored in an amber' or blue glass bottle).

C. Cajal's Reducer.
   Hydroquinonb 2% aqueous 100 ml.
   Neutral formalin 15 ml.
   Sodium sulphite anhydrous 0.5 gm.
   N.B. : This solution should be freshly prepared.

D. Gold chloride 0.2% aqueous.

E. Sodium thiosulphate 5% aqueous.

F. Paracarmine (Mayer).

Procedure

1. Pieces of tissue nb thicker than 3 mm. are fixed from two to eighteen hours in the cobalt nitrate formalin solution, according to the size and nature of the material
2. Wash the tissue quickly in a large volume of distilled water.
3. Immerse in the silver nitrate solution in the dark for thirty-six to forty-eight hours.
4. Wash quickly in a large volume of distilled water. 5. Trim the tissue to a thickness not exceeding 2 mm.
6. Immerse in Solution C (Cajal's Reducer) for two to twenty-four hours in the dark.

   Note : For most soft tissues about four hours will suffice.
7. Wash in several changes of distilled water.
8. Dehydrate, clear and embed in paraffin wax in the usual way.
9. Cut sections up to 8μ in thickness.
10. Fix sections to slides; dewax and pass through descend-ing grades of alcohol to distilled water.
11. Tone sections on slides by immersing in the gold chloride-solution for five to ten minutes.
12. Wash quickly in distilled water.
13. Fix in 5 % sodium thiosulphate (Solution E) for ten to. fifteen minutes.
14. Wash thoroughly in distilled water.
15. Counterstain with paracarmine for about ten minutes.
16. Rinse with 90% alcohol, followed by absolute alcohol.
17. Clear in xylol and mount.

Results

Golgi apparatus stained black while cells are pink or red.

Method 9.9

Gold Chloride Method

Solution required

0.1 % aq. yellow gold chloride solution

5 % sodium thiosulphate solution

Hematoxylin eosin stain solution.

Procedure

1. Bring the section to water and then to distilled water and placed for 5-10 minutes in 0.1% aq. yellow gold chloride solution containing a few drops of glacial acetic acid per 100 ml.
2. Rinse quickly in distilled water.
3. Fix in 5% sodium_ thiosulphate for 5- 10 minutes..
4. Wash in distilled and tap water, stain with hematoxylia and cosine or other desired method.
5. Dehydrate and mount.

Results

Golgi complex-black, Nuclei-blue.

Method 9.10

Osmic Acid Method

Solutions required

(1) Flemmings Fluid;

1% chromic acid	–75 ml
1% Aq. osmium tetraoxide	–20 ml
Sodium chloride	=0.75 gm

(2) Mann's Fixative

1% Aq. osmium tetraoxide	=50 ml
0.75% Aq. sodium chloride	–50 ml

saturated with mercuric chloride.

(3) Osmic acid solution

1% Aq. osmic acid.

Note. All these osmic acid solutions produce very irritating vapours and must be kept tightly sealed and away from the eyes and hands.

Procedure

1. Fix very small pieces of tissue in Flemming's or Mann's fluid for 24 hrs.
2. Wash in running water for 6-12 hrs and transfer to several changes of distilled water.
3. Place in 1 % aq. osmic acid at 37°C for 4 days.
4. Pour off osmic acid and wash in running' tap water for 6–18 hrs.
5. Dehydrate in alcohol or acetone, clear in xylol or tolune and embed in paraffin.

6. Cut sections 4 to 6 microns.
7. Deparaffinise and mount.

Results

Golgi elements-black.

Method 9.11

Cajal Method : (Davenport)

Solution required

1. IN alcoholic Nitric acid solution.
2. Alcoholic silver nitrate solution.

Silver nitrate	=10 gm
Distilled water	=10 ml
When solution is complete then add	
95% alcohol	=90 ml

3. 5% Pyrogallol solution in alcohol

Pyrogallol	-5 gm
95% alcohol	-100 ml

Procedure

1. Fix nervous tissue 2 days or longer in 10-20% formalin.
2. Embed in nitrocellulose, cut sections and place them in 70% or 80% alcohol.
3. Transfer to an approximately IN alcoholic solution of nitric acid made by adding slowly and with stirring 15 ml of conc. HNO_3 to 85 ml of 95% alcohol. Incubate for about 2 hrs. at 37-44°C.
4. Pass the sections individually through 3 changes of 80% alcohol, 1 -2 minutes in each change, wash out the acid.
5. Impregnate at 37-44°C in darkness for 4-24 hrs in the solution of alcoholic silver nitrate.
6. Rinse in 95% alcohol for 3-5 seconds and reduce in 5% Pyrogatled solution prepared in 95% alcohol. (If this solution is to be kept for several days, add 5 ml of formalin to it to retard its oxidation by air. It reduces equally well with or without formalin)
7. Wash well in running water, dehydrate, clear, and mount.

Results

Golgi complex-black.

STUDY OF NUCELIC ACIDS

The nucleic acids -Deoxyribonucleic acid (DNA) and ribosenucleic acid (RNA) are found in all animal and plant tissues, and are usually combined with basic proteins so as to form nucleoproteins. DNA is mainly found in the nucleus, and RNA in the cytoplasm. Hydrolysis of nucleic acids yields the following components : *(a)* phosphate groups, *(b)* five-carbon sugars and (c) nitrogenous bases purines and pyrimidines.

The two enzyme techniques for the digestion of nucleic acids are invaluable as controls for the methods discussed. Dsoxyribonuclease is specific for the removal of DNA ini sections, while not affecting of RNA contents.

When ribonuclease is applied to tissue sections, all the RNA is removed whilst the DNA is unaltered. The digestion methods were introduced by *Bracket* as controls for the Methyl Pysonin method. Great care must be taken that the enzymes are of a high purity, because impure enzymes will remove all nucleic acids. Both ribosenuclease and deoxyri-bosenuclease are expensive reagents. Some of the method are discussed as below

For DNA

Method 9.12

Fealgen Nuclear Reaction Method :
(By Feulgen and Rossenbeek)

Reagents required

(1) Hydrochloric acid
(2) Schiff's reagent
(3) Potassium metabisulphite
(4) Light green, 1 per cent aqueous

Preparation of solutions

(1) n-HCI

Hydrochloric acid, (conc.)	8.5 ml
Distilled water	91.5 ml

(2) Schiff's reagent

(3) Bisulphlte solution

10 per cent Potassium metabisulphite	5 ml
n-Hydrochloric acid	5 ml
Distilled water	90 ml

Sections

All types

Suitable control sections

Pancreas

Procedure

(1)	Bring all sections to water	
(2)	Rinse sections in *n-HCI* at room temperature	1 min
(3)	Place sections in *n*-HCI at 6°C,	
(4)	Rinse sections in *n-HC1* at room temperature	1 min
(5)	Transfer sections to Schiff's reagent	45 min
(6)	Rinse sections in bisutphite, solution (3)	2 min
(7)	Repreat wash in bisulphite, solution (3)	2 min
(8)	Repeat wash in bisulphite, solution (3)	2 min
(9)	Rinse well in distilled water	
(10)	Counterstain if required in 1 per cent Light Green	2 min
(11)	Wash in water	
(12)	Dehydrate through graded alcohols to xylene and mount	

Results

DNA : red-purple Cytoplasm : green

Remarks

(1) The hydrolysis is important, and the correct time for the fixative must be used.

(2) The *n-HCI* should be preheated to 60°C.

Method 9.13

Naphthoic Acid Hydrazine-Feulgen Method

Reagents required

(1) 2-Hydroxy-3-naphthoic acid hydrazine

(2) Absolute alcohol

(3) Acetic acid

(4) Fast blue B

(5) Veronal acetate buffer, pH 7.4

(6) n-Hydrochloric acid

Preparation of solutions

(1) n-Hydrochloric acid

	Hydrochloric acid, (conc.)	8.5 ml
	Distilled water	91.5 ml
(2)	NAH solution	
	2-Hydroxy-3-naphthoic acid hydrazide	50 mg
	Absolute alcohol	47 ml
	Acetic acid (conc.)	3 ml
(3)	Fast Blue B solution	
	Fast Blue B	50 mg
	Veronal acetate buffer, pH 7.4	50 ml
	This solution must ba freshly prepared.	

Sections

All types

Suitable control sections

Pancreas

Method

(1)	Bring all sections to water	
(2)	Rinse briefly in *n*-HCl	
(3)	Place sections in *n*-HCl at 62°C,	
(4)	Rinse sections in *n*-HCl at room temperature	1 min
(5)	Rinse sections in distilled water	1 min
(6)	Rinse sections in 50 per cent alcohol	1 min
(7)	Place sections in NAH solution at room temperature	3-6 h
(8)	Rinse sections in 50 per cent alcohol 10 min	
(9)	Rinse sections in 50 per cent alcohol	10 min
(10)	Rinse sections in 50 per cent alcohol	10 min
(11)	Rinse sections in distilled water	
(12)	Place sections in fresh Fast Blue B solution	3 min
(13)	Dehydrate through graded alcohols to xylene and mount in DPX	

Result

DNA : blue to bluish, purple

Protein material : possibly purplish red

For DNA and RNA

Method 9.14

Methyl Green-Pyronin Method (Modified by Kurwick)

Reagents required

(1) Methyl green
(2) Chloroform
(3) Pyronin Y
(4) 0.1 M Acetate buffer, pH 4.8

Preparation of solutions

(1) Methyl Green

Methyl green	2g
Distilled water	100 mil

Dissolve the Methyl Green in the distilled water by stirring well. Pour the solution into a separating funnel. Add 100 ml chloroform and shake well; poor off contaminated chloroform and repeat until no more violet is extracted (about 6-S washes).

(2) Pyronin Y

Pyronin Y	2 g
Distilled water	100 ml

(3) Staining solution

Methyl Green	7.5 ml
Pyronin Y	12.5 ml
Acetate buffer, pH 4.8	30.0 ml

Sections

All types, freeze dried recommended

Suitable control section

Pancreas

Procedure

(1) Bring all sections to water
(2) Stain in Methyl Green-Pyronin solution, 4-10 min
(3) Blot dry
(4) Rinse rapidly in absolute acetone
(5) Rinse rapidly in 10 per cent acetone in xylene
(6) Rinse rapidly in 50 per cent acetone in xylene
(7) Rinse in xylene
(8) Place sections in fresh xylene and mount in DPX

Results

DNA : green
RNA : red

Remarks

(1) This method can be unreliable, the main causes being that some samples of Pyronin Y will not work satisfactorily, or that the Methyl Green is not pure enough.

(2) The dehydration through acetone and xylene should be rapid.

(3) Triethyl phosphate may be used for dehydration. Instead of acetone.

(4) Methyl Green alone may be used in the manner described above to demonstrate DNA.

Method 9.15

Methyl Green-Pyronin Method
(Modified by Trevan and Sharrock)

Reagents required

(1) 2 per cent Methyl Green.

(2) 5 per cent Pyronin Y

(3) Acetate buffer pH 4.8.

Solution A

2 per cent Methyl Green (chloroform-washed)	10 ml
5 per cent Pyronin Y	17.5 ml
Distilled water	250 ml

Solution B

Acetate buffer pH 4.8

Working solution

Solution A	25 ml.
Solution B	25 ml.

Procedure

(1) Bring sections to water

(2) Rinse in distilled water and blot dry

(3) Stain in working solution 20-30 min

(4) Rinse rapidly in distilled water and blot dry

(5) Dehydrate in acetone

(6) Rinse in acetone xylone 50 : 50, clear in xylene

(7) Mount in DPX

Results

DNA : green, bluish green

RNA : red

Remarks

(1) Fixation should be in neutral fixatives

Method 9.16

Gallocyanin-Chrome Alum Method (By Einarson)

Reagents required

(1) Chrome alum

(2) Gallocyanin

(3) Distilled water

Preparation of solution

Chrome alum	5 g
Distilled water	100 ml
Gallocyanin	150 mg

The chrome alum is dissolved in the, distilled water, the gallocyanin added and the solution slowly heated until it boils. It is allowed to boil for 5 min. When the solution has cooled Io room temperature, the volume is adjusted to 100 ml. The solution is filtered before use.

Sections

All types

Suitable control sections

Pancreas

Procedure

(1) Bring sections down to water

(2) Stain in gallocyanin-chrome alum solution 18-48 h

(3) Wash in tap water

(4) Dehydrate through graded alcohols and mount in DPX

Results

RNA, DNA : blue

Remarks

This method does not distinguish between RNA and DNA but is specific for nucleic acids.

Method 9.17

Acridine Orange Method (Bertalangh and Nagy)

Solution required

(1) Acridine Orange

(2) 0.2 M Phosphate buffer

(3) Calcium chloride
(4) Distilled water
(5) Acetic acid

Preparation of solutions

Acridine orange solution

Acridine Orange	50 mg
Distilled water	40 ml

The pH of the solution is adjusted to 6.0 with phosphate buffer. The volume is then made up to 50 ml.

(1) Phosphate buffer, pH 6.0
See Buffer Tables.

(2) Calcium chloride

Calcium chloride	11 g
Distilled water	50 ml

Sections

Frozen sections Cryostat sections Paraffin sections Freete dried sections Smears	} Not formalin-fixed

Suitable control sections

Pancreas

Procedure

(1)	Bring all sections to distilled water	
(2)	Rinse briefly in 1 per cent acetic acid	15 s.
(3)	Rinse sections in distilled water	15 s.
(4)	Stain in Acridine Orange solution	10 sec.–2 min
(5)	Transfer sections to phosphate buffer, pH 6.0	1 min
(6)	Differentiate sections in calcium chloride solution	20s
(7)	Transfer sections to phosphate buffer, pH 6.0	10s
(8)	Mount sections wet and examine under guorescent microscope	

Results

RNA : red
DNA : light green

Remarks

The authors state that formalin fixation should not be employed

Method 9.18

DNA : Extraction (Brachet)

Reagents required

(1) Deoxyribonuclease

(2) 0.2 M Tris buffer, pH 7.6

Preparation of solution

Extraction of solution

0.2 M Tris buffer, pH 7.6	10 ml
Distilled water	50 ml
Deoxyribonuclease	10 mg

Method

(1) Bring test anfl control sections to water

(2) Place test section in extraction solution, control in tris buffer, pH 7.6 at 37°C, for 4 hrs.

(3) Wash in running tap water

(4) Stain by method (Feulgen method) both sections

Results

Test section : DNA negative Control section : DNA red

Method 9.19

RNA : Extraction (Brachet)

Reagents required

(1) Ribonuclease

(2) Distilled water

Preparation of solution

Ribonuclease	8 ı
Distilled water	10 r.

Procedure

(1) Bring test and control slides to water

(2) Place test slide in ribonuclease solution, control slide· distilled water at 37°C for 1 hr

(3) Wash in distilled water

(4) Apply Method-Methyl Green-Pyronin

Results

Test slide : RNA negative, DNA green

Controlled slide : RNA red, DNA green

Remarks

Nucleic acids may also be extracted by using perchloric acid. This is not as specific as enzyme digestion, but is a reliable technique and may be used on many occasions where extraction is required as considerable reduction is compared with enzyme digestion, the technique is given below.

Method 9.20

Extraction of Nucleic Acids with Perchloric Acid

Reagents required

Perchloric acid
Sodi carbonate

Preparation of solutions

(1)	Perchloric acid	2.5 ml
	Distilled water	47.5 ml
(2)	Perchloric acid	5 ml
	Distilled water	45 ml
(3)	Sodium carbonate	1 g
	Distilled water	100 ml

Procedure

To remove RNA only

(1) Bring sections down to water
(2) Place sections in 10 per cent Perchloric acid (solution 2) at 4°C overnight
(3) Briefly rinse in distilled water
(4) Transfer to the sodium carbonate — 5 min
(5) Wash in tap water
(6) Employ nucleic acid method

Remarks

To remove both RNA and DNA place sections in 5 per cent perchloric acid (solution 1) at 60°C for 30 min at stage (2) then continue method.

STAINING METHODS FOR MAST CELL GRANULES

Method 9.21

Dahlia Acetic Method

Solution required

Distilled water — 100 ml.

Absolute alcohol	50 ml.
Glacial acetic acid	12.5 ml.
Dahlia	10 gm.

Dissolve by heating in a flask, lightly plugged with cotton-wool, on a water bath. Allow to cool; then filter.

Procedure

1. Fix tissues in absolute alcohol and embed in Celloidin.
2. Immerse sections in the staining solution for twelve hours.
3. Differentiate in 95% alcohol.
4. Clear in origanum oil and mount in balsam or in crista-lite.

Results

Granules of mast cells are stained reddish violet.

Method 9.22

Chryoidin Methods

(*For selective staining of mast cell granules*)

Method 1

Solution required

Chrysoidin Y, 0.5% aqueous

Procedure

1. Remove paraffin wax from sections with xylol, and take down to 80% alcohol in the usual way.
2. Stain in the chrysoidin solution for five to ten minutes.
3. Rinse in distilled water.
4. Dehydrate in 96% alcohol and two changes of absolute alcohol.
5. Clear in xylol, and mount in Canada balsam in xylol or in a synthetic mountant such as D.P.X. or Cristalite.

Results

Mast cell granules : deep brown to black in type I cell; brown to yellow in type II cell. Nuclei and other tissue ele-ments; slightly yellowish. Glands in subcutaneous tissues yellow.

Note. The author states that in using simple aqueous solu-tions of various basic dyes in investigating the staining of tissue containing acid polysaccharides, it was found that only chry-soidin stained the mast cell granules selectively.

Method 2

Chrysoidin-eosin-light green-haematoxylin

Solution required

A.	Alum haematoxylin (Carazzi) Haematoxylin	0.1 gm.
	Potash alum	5 gm.
	Distilled water	80 ml.
	Glycerine	20 ml.
	Potassium iodate	0.02 gm.
B.	Light green SF, 1 % aqueous	1 ml.
	Distilled water	49 ml.
	Or	
C.	Eosin, 1 % aqueous	0.5 ml.
	Tap water	49.5 ml.

Note. Solution C contains 1 : 10,000 of eosin and this may be further diluted, to as 1 : 50,000, with tap water, if desired.

D. Chrysoidin Y, 0.5% aqueou&

Procedure

1. Take sections down to 80% alcohol as usual.
2. Stain in the alum haematoxylin for five to ten minutes.
3. Wash in tap water.
4. Stain for three minutes in either solution B or C.
5. Rinse in water.
6. Stain in the chrysoidin solution for five to ten minutes.
7. Rinse in water.
8. Dehydrate with 96% alcohol and two changes of absolute alcohol; then clear and mount.

Results

Mast cell granules : brown. Nuclei : greyish. Epidermis slightly yellow, Glands : yellow. Connective tissue : green or red.

Method 3

Chrysoidin-periodic acid-Schiff

Solution required

A. Periodic acid, 0.7% acqueous

B.S chiff's reagent

C. Sodium bisulphite, 0.52% aqueous.

D. (Optional) Alum haematoxylin (Carazzi)

E.C hrysoidin Y, 0.5% aqueous

Procedure

1. Immerse in solution A for ten minutes.

2. Rinse in water.
3. Immerse in the Schiff's reagent for fifteen to thirty minutes.
4. Immerse for one minute in each of three changes of the sodium bisulphite solution.
5. Wash in running water for ten minutes.
6. (Optional) Stain in the haematoxylin solution for five to ten minutes.
7. Wash in water.
8. Stain in the chrysoidin solution for ten minutes.
9. Rinse in water.
10. Dehydrate with 96% alcohol and two changes of absolute alcohol.
11. Clear in xylol and mount in Canada balsam in xylol or in a neutral synthetic medium.

Results

Mast cell granules : brown. Nuclei : grey (weakly yellowish). Epidermis : slightly yellow. Glands : yellowish. Connective tissue : intense purple.

In type II cells of abdominal fluids, the granule or its halo of the outer layer is stained weakly with the periodic acid-· Schiff procedures.

STAINING METHOD FOR NEGRI BODIES

Method 9.23

Carbol Aniline Fuchsin Method

Solution required

A.	Basic fuchsin	0.5 gm.
	Distilled water	80 ml.
	Absolute alcohol	20 ml.
	Aniline oil	1 ml.
	Phenol	1 gm.
B.	Methylene Blue (Loeffler).	

Procedure

1. Fix tissues in Zenker's Fluid for twenty-four hours; wash in running water for two or three hours; dehydrate; clear; embed in paraffin wax in the usual manner.
2. Sections, 4 to 5μ in thickness, are stained from ten to thirty

minutes in Solution A; then washed with distilled water.

3. Stein with Methylene Blue (Loffier) for fifteen to sixty seconds; then wash with water.
4. Dehydrate and differentiate for a few seconds in absolute alcohol; then clear in xylol and mount.

Results

Negri bodies are stained crimson against a blue background.

Note. The method is stated to be excellent for Borrel bodies also.

DETECTION OF SEX CHROMATIN

Method 9.24

Cresyl Fast Violet Method

Solution required

A.	Absolute alcohol	equal
	Ether	volumes
B.	Cresyl fast violet CNS	1 gm.
	Alcohol, 50%	100 ml.

Procedure

1. Take scrapings, from the inside of the check, with a spatula.
2. Spread the scraping on a slide.
3. Fix immediately in the alcohol-ether mixture, for at least fifteen minutes.

Note. The author (Moore) states that the usual causes of poor smears are : (a) not scraping firmly enough, with the result that the number of cells obtained is insufficient, and (b) allowing the smear to dry before fixation, resulting in poor nuclear detail.

4. Immerse the preparation in 70% alcohol for two minutes.
5. Immerse in 50% alcohol for two minutes.
6. Wash for one minute in each of two changes of distilled water.
7. Stain in the cresyl fast violet solution for seven to eight minutes.
8. Differentiate by dipping the slide quickly about five to seven times in 95% alcohol.
9. Dehydrate in absolute alcohol for one minute.
10. Mount in clearmount or michrome mountant, or pro-ceed as follows
11. Wash in xylol.

12. Mount in clearmount or michrome mountant or D.P.X. or, if these are not available, in neutral Canada balsam in xylol, although this is liable to cause fading of the stain.
13. Examine under the oil immersion objective with a strong. source of light.

Results

The sex chromatin is visible only in the nucleus from the female cells.

10

SPECIFIC STAINING FOR VARIOUS TISSUES

Certain specific stains are used for identifying and studying connective tissues such as muscle fibres, blood cells etc. The three main fibrous connective tissues are *collagen, reticulin* and *elastin*. These fibres have certain physical and chemical properties which enable them to be stained individually.

Collagen is a basic glycoprotein containing high propor-tions of glycine and protine. *Hydroxylysine,* an animo acid is also found in collagen in small quantity. In the electron, microscope, collegen fibres present a triple helices joined end to end and side to side by hydrogen bonds by covalent bridges and electrostatic attractions. Under light microscope, they seem to be bundles of these fibres.

The term "reticulin includes the basement membranes of epithelia and blood capillaries. Both these types contain more carbohydrates then collagen. The third type, elastic fibres are made of elastin. The elastin is a hydrophobic protein, rich in glycine, alamine and value.

The staining techniques for connective tissues fall into three categories those based on the use of mixture of anionic dyes to give different colours to collogen and cytoplasm, the methods for reticulin and methods for elastin. Some of the individual methods for all the three types are given as below.

STAINING METHODS FOR COLLAGEN

Many techniques are available for staining with mixtures of,dyes.

Collagen and cytoplasm are coloured differently but other elements such as cartilage, fibrin and secretory granules aquire characteristic colours.

Method 10.1

Iron-haemotoxylin and van Gieson Method

Solutions required

1. Weigert's iron-haematoxylin

   Solution A. Haematoxylin

(C.I. 75290) :	5 g
95% ethanol	500 ml

   Keeps for a few years

   Solution B. Ferric chloride

($FeC1_s.6H_sO$) :	5.8 g
Water :	495 ml Concentrated
hydrochloric acid :	5 ml

   Keeps indefinitely

Working solution

Mix equal volumes of A and B. Put A in the staining jar or tank first for more rapid mixing. The mixture should be made just before using, but can be kept for about 2 weeks at 4°C.

2. Van Gieson's solution

   Acid fuchsine

(C.I. 42685) :	0.5 g	Keeps indefinitely

   Saturated aqueous

   picric acid

See *Note* 2 below.

Procedure

1. De-wax and hydrate paraffin sections.
2. Stain in working solution of Weigert's haematoxylin for 5 minutes (10 minutes if the solution is more than a few days old).
3. Wash in running tap water.
4. Stain in van. Gieson's solution, 2-5 min. The time is not critical.
S. Wash briefly in running tap water. This also differentiates the acid fuchsine.
6. Dehydrate rapidly three changes of 100% ethanol. This step also differentiates the picric acid.

7. Clear in xylene and mount in a resinous medium.

Result

Nuclei-black or brown; collagen-red; cytoplasm (especially smooth and striated muscle), keratin and erythro-cytes-yellow.

Method 10.2

Cason's Trichrome Staining Method

Solution required

A Weigert's iron-haematoxylin (working solution).

See above.

B. Cason's trichrome solution

Water :	200 ml 1	
Dissolve in order stated		
Phosphotungstic acid	1 g	Keeps for
Orange G (C.I. 16230)	2 g	1-2 years
Aniline blue WS (C.I. 42755)	1 g	
Acid fuchsine (C.I. 42685)	3 g	

Procedure

1. De-wax and hydrate sections.
2. Stain in Weigert's haematoxylin, 5 minutes.
3. Wash in running tap water, 2 minutes.
4. Immerse in Cason's trichrome solution, 5 minutes.
5. Wash in running water, 3-5 seconds.
6. Blot slides dry with filter paper.
7. Dehydrate rapidly in three changes of 100% ethanol.
8. Clear in xylene and cover, using a resinous medium.

Results

Collagen-blue; cytoplasm, muscle-red; keratin, erythrocytes -orange; nuclei-brown (but sometimes blue). The pre-staining with iron-haematoxylin makes the trichrome colours a little unpredictable with some material. If the nuclei stain with iron-haematoxylin is omitted, most nuclei are coloured red; others may be blue or unstained.

Method 10.3

Orcein-Picro Fuchsin Method

Solution required

A. Orcein (Unna) 1% in 80% alcohol 100 ml

Hydrochloric acid, conc. 1 ml

B. Picro-fuchsin (Van Gieson).

Procedure

1. Sections are mounted on slides and brought down to 70% alcohol in the usual manner. If tissues have been fixed in a fluid containing mercury, the mercurial precipitate is removed by the standard technique.
2. Immerse is orcein solution (recipe as above) for half an hour or longer if necessary : then rinse in acid alcohol.
3. Rinse in 70% alcohol; then in water.
4. Stain with picro fuchsin (Van Gieson) for three to five minutes.
5. Rinse rapidy (not more than a few seconds) in water.
6. Dehydrate rapidly; clear, then mount.

Results

Collagen fibres are stained red; elastic fibres, brown; erythrocytes, epithelia, muscle, etc, yellow.

Method 10.4

Masson's Trichrome Staining Method

Solution required

A. Iron alum 5% aqueous.

B. Regaud's haematoxylin solution.

C. Picric acid, saturated in 95% alcohol 20 ml.
Alcohol 95% 10 ml.

D. Ponceau fuchsin.

E. Phosphomolybdic acid 1 % aqueous.

F. Aniline Blue 5% in 2% acetic acid.

Procedure

1. Fix pieces of tissue in Bouin's fluid for three days or in Regaud's fluid for one day.
2. Wash in running water; dehydrate; clear and embed in paraffin wax as usual.
3. Section 5μ in thicknes are fixed to slides; de-waxed and passed through descending grades of alcohol down to distilled water in the usual manner.
4. Mordant in Solution A for five minutes at 45°C to 50°C.

5. Wash well in distilled water.
6. Stain for five minutes in Regaud's haematoxylin at 45°C to *50* C.
7. Rinse in distilled water.
8. Differentiate in picric alcohol (Solution C above) con-trolling by examination under the microscope, while the preparation is still wet.
9. Wash in running tap water for a minute or so.
10. Stain for five minutes in the Ponceau fuchsin solution.
11. Rinse in distilled water.
12. Differentiate in the phosphomolybdic acid solution for five minutes.
13. Add 0.5 ml of the acetic aniline blue (Solution F above) to the phosphomolybdic acid on the slide and mix by rocking the slide gently. Allow this mixture to act for five minutes.
14. Pour off excess liquid and rinse in distilled water.
15. Immerse in phosphomolybdic acid solution again, for five minutes.
16. Transfer to 1 % acetic acid and leave therein for five minutes.
17. Wash in distilled water.
18. Dehydrate in 95% alcohol, followed by absolute alcohol; clear in xylol; mount.

Results

Collagen, deep blue. Neuroglia fibrils, red. Nuclei black. Argentaffln granules, black or red.

Method 10.5

Saffron-Erythrosin Method

Solution required

A.	Saffron	2 gm
	Distilled water	100 ml

Boil gently for an hour; allow to cool : then filter; and 1 ml of 40% formaldehyde and 1 ml of 5% tannic acid to the filtrate.

Note. Saffron solution deteriorates after a few weeks, and it is best to prepare the solution in small quantities, as required.

B: Delaficid or Ehrlich haematoxylin.

C. Erythrosin, 1 % aqueous.

Procedure

1. Fix small pieces of tissue in Bouin, Zenker-formaldehyde or in mercuric-formaldehyde.
2. Wash; dehydrate; embed.
3. Sections are stained for five or ten minutes with Delafield or Ehrlich haematoxylin; rinse **in** water.
4. Blue in tap water in the usual manner **or in 1 % sodium** phosphate (Na_2HPO_4).
5. Stain for two to five minutes in 1 % aqueous crythrosin. 6. Rinse quickly with water.
7. Differentiate with 70% alcohol for a few seconds, controlling under the microscope, until the collagen fibres are nearly colourless.
8. Rinse in water; stain for five minutes in saffron solution prepared as above; rinse with water.
9. Wash rapidly first with 70% alcohol then with absolute alcohol; clear in xylol and mount.

Results

Nuclei-blue; cytoplasm-varying shades of red; Muscles-pinta, elastic fibres—pink, collagen-yellow.

Method 10.6

Aniline Blue-Orange G-Acid Fnschsin Method

Solutions required

A. Acid fuchsin 0.5% aqueous.

B. Aniline Blue-Orange G.

Procedure

Tissues are fixed in Zenker and embedded in paraffin wax, Celloidin of L.V.N.

1. Mount sections on slides and bring down to 90% alcohol; then treat with iodine in the usual way to remove mercuric deposits.
2. Bring down to distilled water and stain for one to ten minutes in Solution A; then without washing.
3. Stain for twenty minutes to one hour or longer in Aniline Blue-Orange G; then remove excess stain with several changes of 95% alcohol.
4. Dehydrate with absolute alcohol; clear in xylol and mount Cristalite.

Note. If Celloidin or L.V.N. sections are used the staining time may be shortened and 95% alcohol should be used for decolourising and dehydration; terpineol for clearing.

Results

Collagenous fibrils : intense blue. Ground substances of cartilage, bone, mucus, amyloid : varying shades of blue. Nuclei, myoglia, neuroglia fibrils, axis cylinders, fibrin, nucleoli red. Blood corpuscles and myelin : yellow. Elastic fibrils pale pink or pale yellow, or unstained; fibriloglia : red or unstained.

Note. By omitting the acid fuchsin the collagenous fibres are more sharply defined.

Method 10.7

Mallory Heidenhain Staining Method

Solution required

Phosphotungstic acid crystals A.R.	1 gm
Orange **G**	2 gm
Aniline blue, water soluble	1 gm
Acid fuchsia	3 gm
Distilled water	200 ml

Procedure

1. Fix pieces of tissue in Zenker-formol for preference, although Bouin's fluid, formal in and alcohol have been used with success.
2. Embed in paraffin wax and cut sections 6,u in thickness.
3. Fix sections to slides and remove wax with xylol.
4. Pass through descending grades of alcohol and if Zenker formol has been used as the fixative, treat with iodine and. sodium thiosuiphate as usual to remove mercurial precipitate.
5. Take down to tap water.
6. Immerse for five minutes in the staining solution.
7. Wash in running tap water for three to five minutes.
8. Dehydrate rapidly through the usual graded alcohols.
9. Clear in xylol and mount.

Results

Appear to be the same as those listed by Mallory, i.e., collageneous fibrils, intense blue. Ground-substance of cartilage and bone, mucus, amyloid, and certain other hyaline substance are stained in varying shades of blue. Nuclei, fibro-glia, myoglia and neuroglia fibrils, nucleoli, axis cylinders and fibrin are stained red. Erythrocytes and myelin, yellow. Elastic= fibrils are stained pale pink or yellow.

Method 10.8

Haematoxylin-Biebrich Scarlet Picro Aniline Blue Method

Solution required

A.	Haematoxylin (Weigert) A.	
B.	Haematoxylin (Weigert) B.	
C.	Biebrich scarlet, 0.2% aqueous	100 ml
	Glacial acetic acid	1 ml
D.	Picric acid, saturated aqueous	100 ml
	Aniline blue, water soluble	0.1 gm
E.	Acetic acid 1 % aqueous.	

Procedure

1. Tissues should be fixed in 10% formalin and paraffin sections employed.
2. Stain for five minutes in a freshly prepared mixture consisting of equal parts of Weigert's Haematoxylin A and B.
3. Wash in tap water.
4. Stain for three to five minutes in the acetic Ponceau, S (Solution C).
5. Rinse in distilled water.
6. Stain for three to five minutes in the picro aniline blue (Solution D).
7. Wash for three or four minutes in I % acetic acid solution.
8. Dehydrate in ascending strengths of alcohol and clear in xylol in the usual manner.
9. Mount in acid balsam.

Results

Connective tissue, glomerular basement membrane and reticulum: blue. Muscle and plasma : pink. Erythrocytes bright red.

Method 10.9

Mallory's Phosphotungstic Acid Haematoxylin

Solution required

Haematoxylin 10% in absolute alcohol (ripened for three months or longer)	1 ml

Phosphotungstic acid	2 gm
Distilled water	100 ml

Note. If ripened haematoxylin solution is not available, the following artificially ripened stain should be used

Haematoxylin (dry) 0.1 gm, phosphotungstic acid 2 gm, distilled water 100 ml, potassium permanganate 1 % aqueous 1.77 ml.

Procedure

1. Fix in Zenker. Embed in paraffin wax.
2. Bring sections down to distilled water.
3. Treat with iodine to remove mercuric precipitate.
4. Remove iodine with 0.5% aqueous sodium hyposulphite.
5. Wash thoroughly in running water.
6. Immerse for five to ten minutes in 0.25% potass, permanganate; then wash in tap water.
7. Immerse for ten to twenty minutes in 5% oxalic acid; then wash thoroughly with tap water.
8. Stain twelve to twenty-four hours in haematoxylin solution, prepared as above.
9. Wash in tap water; dehydrate with 95% and absolute alcohol.
10. Clear in xylol and mount.

Results

Nuclei, centrioles, achromatic spindles, fibroglia, myoglia, neuroglia fibrils, fibrin, contractile elements of striated muscle blue. Collagen, reticulum, ground substances of cartilage and bone : yellowish to browuish red. Coarse elastic fibrils : faint purple.

METHODS FOR RETICULUN

Method 10.10

Gordon and Sweets 1939 Method

Solution required

(A) Acid permanganate

Potassium permanganate ($KMnO_4$)	1.0 g.
Water	95 ml
3% aqueous H_2SO_4	5 ml

Prepare just before use from a stock 6% aqueous-$KMnO_4$. The addition of H_2SO_4 is not necessary.

(B) 1% oxalic acid

Oxalic acid ($HOOC. COOH.2H_2O$)	5 g
Water	to 500 ml

Keeps indefinitely

(C) Iron alum

Iron alum ($NH_4Fe(SO_4)_2.12H_2O$)	10 g
Water	to 500 ml

Prepare on the day it is to be used. Alternatively, use 4% aqueous ferric chloride ($FeCl_8$. $6H_20$), which keeps indefinitely

(D) Ammoniacal silver solution

Stock Solutions : 1. 10% aqueous silver nitrate.
2. Ammonium hydroxide (28% NH_8).
3. 4% aqueous sodium hydroxide.

Working Solution

Add ammonium hydroxide drop by drop to 10 ml of 10% $AgNO_3$ until the brown precipitate of Ag_2O is almost (not quite) resolved. Add 7.5 ml of 4% NaOH, followed by a few more drops of ammonium hydroxide, until the realy formed precipitate is just dissolved. Be careful not to add too much ammonia. Swirl the solution by a few seconds after adding each drop, since disolution of the precipitate is not quite instantaneous. Make up to 100 ml with water. This solution should be made just before use and discarded afterwards by washing' it down the sink with plenty of water. Ammoniacal silver solutions decompose on evaporation to form explosive "*fulminating silver*", a mixture of silver amide and silver nitride.

(E) Reducer

Neutralised formalin (40% HCHO which has stood over marble) :	10 ml	Prepare just before using
Water	90 ml	

(F) Yellow gold chloride

Sodium tetrachloroaurate ($NaAuCl_4.2H_sO$)	1 g	Keeps for several months
Water	500 ml	

This solution may be re-used repeatedly

(G) Sodium thiosulphate Sodium thiosulphale

($Na_2S_2O_8$. $5H_2O$) :	25 g	Keeps for several months
Water :	to 500 ml	

Procedure

1. De-wax and hydrate paraffin sections.
2. Oxidise for I min in acid permanganate (solution A).
3. Wash in water.
4. Immerse in 1 % oxalic acid (solution B) until the sections are white. Usually about 30 s.
5. Wash in water (three changes).
6. Treat with iron alum (solution C), 10 min.
7. Wash in water (three changes).
8. Immerse slides in the ammoniacal silver solution (solution D) 5-10 s.
9. Rinse in water (once only).
10. Place in formaldehyde reducer (solution E), 30 s.
11. Wash in water (three changes).
12. Tone in 0.2% yellow gold chloride (solution F), 2 m.
13. Wash in water (two changes).
14. Immerse in sodium thiosulphate (solution G), 3 min,
15. Wash in water (three changes).
16. Dehydrate through graded alcohols, clear in xylene and cover.

Result

Reticulin- black. Other elements in shades of grey.

METHODS FOR ELASTIN

Method 10.11

Weigert Elastin Staining Method

Preparation of the Staining Solution

Triturate gm of Weigert elastin stain and 5 gm clean, dry silver sand with 100 ml absolute alcohol and 2 ml pure hydro-chloric acid until all the stain has gone into solution; then. filter.

Note. The staining solution deteriorates after two or three weeks.

The nuclei may be stained with Orth's lithium carmine prior to the following procedure if no other counterstain is desired.

Procedure

1. Sections are brought down to 90% alcohol and stained one half to twelve hours according to depth of staining desired. The slides should be stained in a jar or in a Petri dish, sections face downwards, to prevent a deposit forming on the sections.

2. Wash off excess stain with 95% alcohol, and if necessary differentiate in acid alcohol for a few minutes.
3. Wash quickly with 70% alcohol; then thoroughly with water.
4. Counterstain with Van Gieson, Ehrlich haematoxylin or Safranin for about five minutes.
5. Differentiate, if necessary, in 95% alcohol.
6. Dehydrate; clear in xylol and mount.

Note. If Celloidin or L.V.N. sections are used clear in origanum oil or in terpineol after 95 % alcohol.

Results

Elastic fibres, dark blue or black, Nuclei, brilliant red (if Orth's carmine is used) or bluish black (with haematoxylin).

Collagen, pink to red : other tissue elements, yellow (if Van Gieson is used).

Method 10.12

Elastin-Trichrome Staining Method

Solutions required

A. Weigert's elastin stain

Weigert's elastin stain powder	1 gm
Hydrochloric acid, conc., pure	2 ml
Absolute alcohol	100 ml

Dissolve the stain by boiling for two minutes in a flask, plugged lightly with cotton-wool, on a water bath. Allow to cool; then filter; make the volume up to 100 ml with absolute alcohol; then add the acid. Alternatively, the solution may be prepared as described.

Note. This solution deteriorates after three or four weeks.

B. Ehrlich haematoxylin.

C. *Ponceau-acid fuchsin (Masson) :*

Acid fuchsin	0.3 gm
Ponceau de xylidine	0.7 gm
Distilled water	100 ml
Glacial acetic acid	1 ml

D. Phosphotungstic acid 3% aqueous

E. Light Green 1 % aqueous.

Procedure

1. Paraffin sections are mounted on slides and brought down to distilled water in the usual manner; then immersed in Weigert's

elastin stain in a staining jar for one hour.

2. Wash rapidly in acid alcohol; then dehydrate and differentiate in absolute alcohol until the sections appear only faintly red.
3. Immerse in 70%. alcohol, followed by distilled water.
4. Stain in Ehrlich haematoxylin for eight to ten minutes; then differentiate in water for five minutes.
5. Stain in Ponceau-acid fuchsin for five minutes.
6. Wash thoroughly in 3% phosphotungstic acid; then immerse in the phosphotungstic acid for ten minutes.
7. Wash thoroughly in distilled water; then stain with Light Green for two to five minutes; then without washing.
8. Flood the preparation with 1% acetic acid and allow it to act for three minutes; pour off cxcess; then without wash-ing.
9. Dehydrate; clear; mount in D.P.X.

Results

Elastic tissue stained blue-black; smooth muscle, red collagen, green.

Method 10.13

Congo Red-Aniline Blue-Orange G Method

Solutions required

A.	Aluminium chloride, 2% aqueous.	
B.	Congo Red	2 gm
	Sodium citrate	2.5 gm
	Glycerin	1 ml
	Distilled water	97 ml
C.	Aniline Blue, aqueous	1.5 gm.
	Orange G	2.25 gm
	Resorcinol	3 gm
	Phosphomolybdic acid	1 %
	aqueous	100 ml

Procedure

Tissues should be fixed in 10% formalin, and frozen sec tions should be employed.

1. Wash sections in water; then immerse them in Solution; A for ten minutes.
2. Wash with water and drain; then stain in the Congo Red solution for ten minutes.

3. Wash with tap water; then plunge the slide into a dish of tap water and agitate it there for ten seconds.
4. Wash again with tap water; then stain from five to ten minutes in the Aniline Blue-Orange G solution (solution C above).
5. Rinse carefully in tap water; drain well and blot.
6. Dehydrate in absolute alcohol; clear in origanum oil; wash in xylol and mount.

Results

Elastic fibres : bright red. Fibrin : dark blue. Erythrocytes yellowish orange.

Method 10.14

Orcinol-New Fuchsin Method

Solutions required

A.	New fuchsin	2 gm
	Orcinol	4 gm
	Distilled water	200 ml
	Boil for five minutes; then add	
	Ferric chloride ($FeCl_3$), 15%	25 ml
	and boil for a further period of five minutes.	

Allow the solution to stand until cold; then collect the precipitate by filtration : washing and drying is stated to be unnecessary.

Dissolve the precipitated orcinol-new fuchsin in 100 ml of 95% alcohol and use this as the elastin stain.

Procedure

1. Material may be fixed in 10% formalin or saturated aqueous mercuric chloride or Zenker, etc.
2. Fix sections to slides, dewax and pass through absolute alcohol as usual.
3. If a mercury containing fixative has been used, treat sections for the removal of mercurial precipitate by the standard method, afterwards rinsing with 90% alcohol.
4. Stain in the orcinol new fuchsin for fifteen minutes at 37°C.
5. Differentiate for five minutes in each of three changes of 70% alcohol.
6. Dehydrate in absolute alcohol; clear in xylol, and mount in DPX, or Cristalite, or Clearmount, etc.

Results

Elastic fibres : deep violet. Collagen : unstained.

Notes

(a) The authors reported that they used the stain on human skin and aorta, and the following tissues of the mouse : liver, kidney, spleen, stomach, doudenum, colon, pancreas, heart, testes, seminal vesicles, ovary, uterus, aorta, pituitary, salivary gland, bone, striated muscle, and thyroid; and in all cases only elastic tissue was stained.

Method 10.15

Orcein-Anifime Blue Orange G.

Solutions required

A.	Orcein	1 gm
	Alcohol 70%	100 ml
	Hydroch loaric acid, conc.	0.6 ml
B.	Alcohol 50%	49 ml
	Hydroch laric acid, conc.	0.5 ml
C.	Mallory's Aniline Blue-Orange G.	
	Aniline blue, aqueous	0.5 gm
	Orange G	2 gm
	Phosphornolybdic acid 1 %	100 ml
D.	Orange G 0.1 % in absolute alcohol.	

Procedure

1. Fix material in Bouin and embed in paraffin wax.
2. Sections, about 8μ in thickness are fixed to slides, dewaxed with xylol and passed through the usual descending,grades of alcohol tv distilled water.
3. Stain for one and a half hours in the orcein solution in a closed staining jar.
4. Differentiate with solution B, controlling under the microscope, unti 1 most of the pink is extracted from the sections.

   Note. The duration of the differentiation will vary accord-ing to the nature of the material and to the thickness of the sections.
5. Wash thoroughly' with running tap water.
6. Wash with distilled water.
7. Immerse in solution C diluted with an equal volume of distilled water, for one to three minutes.
8. Rinse with 95% alcohol.
9. Rinse with two lots of solution D.

10. Rinse quickly with absolute alcohol.
11. Clear in xylol and mount.

Results

Elastic fibres : red. Collagen : blue. Muscle fibres : pale orange to dirty yellow. Cytoplasm : varying shades of yellow. Erythrocytes : golden yellow. Kerantinised material : bright yellow.

SOME GENERAL STAINING METHODS FOR CONNECTIVE TISSUE

Method 10.16

Haematoxylin-Picro Fuchsia Method

Solutions required

A.	Distilled water	47.5 ml
	Ferric chloride, hydrated 4% aqueous	2 ml
	Hamaetoxylin 10% in absolute alcohol	0.4 ml
B.	Picric acid, saturated, aqueous	20 ml
	Acid fuchsia 1 % aqueous	0.5 ml
C.	Picric acid, saturated in absolute alcohol.	

Procedure

1. Tissues are fixed in Bouin and embedded in paraffin wax.
2. Sections about 8μ in thickness are fixed to slides, dewaxed with xylol and taken through the usual decending grades of alcohol to distilled water.
3. Stain for two to three minutes in solution A.
4. Differentiate and counterstain for about ten to fifteen seconds in solution B, controlling under the microscope, until only the nuclei are stained a greyish colour with the haematoxylin.
5. Rinse immediately in distilled water.
6. Dehydrate by dripping solution C onto the slide.
7. Clear with Terpineol.
8. Mount directly with Michrome mountant, or rinse with xylol, then mount with Clearmount or Cristalite.

Results

Chromatin : black to grey. Muscle : yellow. Connective tissue : red. Keratinised regions : bright yellow. Cytoplasm yellow.

Method 10.17

Methyl Violet-Pyronin-Orange G Method

Solution required

A.	Methyl violet 6B (Jensen) 1% aqueous	25 ml
	Pyronin B 10% aqueous	10 ml
	Distilled water	65 ml
B.	Acetone	100 nil
	Orange G aqueous 2%	about 10 ml

Add orange G solution drop by drop to the acetone, with shaking, until the flocculent precipitate formed just redissolves with further addition of orange G solution.

Procedure

1. Small pieces of tissue are fixed in acetic-alcohol or in mercuric chloride.
2. Wash; dehydrate; clear, embed in paraffin wax.
3. If mercuric chloride has been used for fixation treat sections for the removal of mercuric precipitate by the standard method.
4. Take sections down to distilled water.
5. Immerse for two minutes in the methyl violet pyronin (Solution A above).
6. Pour off excess stain and carefully wipe the slide dry.
7. Flood the preparation with acetone-orange G solution.
8. Pour off after a few seconds.
9. Flood the preparation with a fresh lot of *acetone-orange G* solution and pour off after a few seconds.
10. Wash quickly in pure acetone.
11. Rinse with two lots of xylol.
12. Mount in balsam.

Results

Cytoplasm, red; chromatin, violet : keratin, violet; connect tive tissue, yellow.

Method 10.18

Nephthol Green-B-Haematoxylin Method

Solutions required

A. Weigert's haematoxylin, A.
B. Weigert's haematoxylin, B.
C. Eosin, yellowish, 1% in tap water.
D. Ferric chloride, hydrated 10%.

E. Naphthol Green B, 1% aqueous.

F. Equal volumes of acetone and xylol.

Procedure

1. Paraffin sections :e re mounted on the slide and brought down to distilled water in the usual manner.
2. Stain for six minutes in a freshly prepared mixture consisting of equal volumes of Weigert's haematoxylin A and B.
3. Wash thoroughly in tap water; then stain for three minutes in the eosin solution.
4. Wash in tap water; then immerse in the ferric chloride solution for five minutes.
5. Rinse well in distilled water; then stain for five minutes in the naphthol green solution.
6. Differentiate for two or three minutes in 1% acetic acid.
7. Drain well; then dehydrate with acetone, afterwards clearing in acetone-xylol (as above); then mount.

Results

Connective tisse, green; muscle and cytoplasm, pink.

Method 10.19

Aniline-Crystal Violet-Lithium Carmine-Iodine Method

Solutions required

A.	Lithium carmine	
B.	Crystal violet	1 gm
	Aniline ail	3 ml
	Absolute alcohol	10 ml
	Dissolve and filter.	
C.	Crystal violet 2% aqueous.	
D.	Solution B	3 ml
	Solution C	27 ml

This mixture should be prepared immediately before use.

E.	Gram's iodine.	
F.	Aniline oil	1 volume
	Xylol	1 volume

Procedure

1. Fix material in absolute alcohol, Carnoy or alcohol-formalin and embed in paraffin wax.

2. Fix sections to slides; de-wax and pass through descending grades of alcohol down to distilled water in the usual way.
3. Stain in the lithium carmine solution for two to five minutes.
4. Wash thoroughly in distilled water.
5. Immerse in Solution D for five to ten minutes.
6. Rinse in distilled water; drain well and blot carefully.
7. Cover with Gram's iodine solution and allow the stain to act for five to ten minutes.
8. Pour off the excess iodine solution and blot carefully with filter paper.
9. Differentiate with the the aniline xylol solution until no more purple coloration comes out.
10. Drain, and blot carefully.
11. Rinse with several changes of xylol.
12. Mount in balsam or in cristalite.

Results

Fibrin and Gram-positive organisms are blue to blue-black-while nuclei are red.

Method 10.20

Alizarin Red-S Method (Dawson's Method)

Solutions required

A.	Potassium hydroxide 1% aqueous	
B.	Alizarin Red, S	0.1 gm
	Potassium hydroxide	10 gm
	Distilled water	1 litre
C.	Mall's solution	
	Glycerin	20 ml
	Distilled water	79 ml
	Potass, hydroxide	1 gm

Procedure

1. Whole specimens are fixed in 95% alcohol for at least three days.
2. Transfer to acetone and leave for several days to dissolve out the fats which would otherwise stain intensely and obscure the view of the bony structures.
3. Wash well with 95% alcohol; then immerse in 95% alcohol

for twenty-four hours.

4. Immerse in Solution A from one to seven days, according to the size of the specimen, until the bones are clearly visible through the muscle.
5. Transfer to Solution B until the bones are stained the desired depth of colour; this takes from one to seven days, and the solution should be changed on the fourth day.
6. Clear in Solution C until no more colour comes out.
7. Pass into a mixture of equal parts of glycerin and water, .and continue through increasing strength of glycerin.
8. Store in pure glycerin.

Results

Bones are stained red ; soft tissue, transparent and unstained.

Method 10.21

Benda's Method (for Nervous Tissues)

Solutions required

A.	Nitric acid, conc.	1 volume
	Distilled water	10 volumes
B.	Potassium dichromate 2%	
C.	Chromic acid 1 D. Iron alum 4%	
E.	Alizarin Red, S, saturated in	
	Absolute alcohol	1 ml
	Distilled water	90 ml
F.	Toluidine Blue 0.1 % aqueous	

Procedure

1. Material is fixed in 90-95% alcohol for at least two days.
2. Pieces, which must not be thicker than 0.5 cm, are immersed in Solution A for twenty-four- hours.
3. Transfer to Solution B for twenty-four hours.
4. Transfer to Solution C for forty-eight hours ; then wash in water for twenty-four hours.
5. Dehydrate in the usual manner.
6. Clear in beechwood creosote for twenty-four hours ; then in benzol for twenty-four hour
7. Embed in paraffin wax via four graded mixtures of paraffin wax and benzol, the firm at room temperature, the second at 38°C, the third at 42°C and the fourth at 45°C.

8. Mount sections; slides to bring down to distilled water, mordant sections on yes with Solution D for twenty-four hours, then wash thoroughly in water.
9. Stain for two hours with Solution E; then rinse in tap
10. Flood slides with Solution F and warm gently until vapour is given off; or stain at room temperature for 2 + hours.
11. Rinse in 1% acetic acid ; then dry by blotting carefully.
12. Pass through absolute alcohol; then differentiate for about ten minutes in beechwood creosote; dry by bloting. carefully, wash with xylol, and mount.

Result

Soft tissue-Transparent; Osseous tissue-deep blue.

Method 10.22

Ammoniacal Silver Carbonate Method (for tumor cells)

Solutions required

A. Pyridin, pure — 2 volumes
   Glycerin, pure — 1 volume

B. Ammoniacal Silver Carbonate

Ammonia solution is added drop by drop to 10 ml silver nitrate 10.2 % until the precipitate formed is almost redissolved, leaving a slightly opalescent solution to which is then added 10 ml sodium carbonate 3.1 % solution and sufficient distilled water to make the volume up to 100 ml.

C. Reducing solution

Sodium carbonate anhydrous	1 gm
Formalin	1 ml
Distilled water	103 ml

D. Brown gold chloride 0.2% aqueous.

E. Intensifying solution

Oxalic acid 2% aqueous	100 ml
Formalin	1 ml

F. Sodium hyposulphite 10% aqueous.

Procedure

The material is fixed in 10% formalin or in Bonin and embedded in paraffin wax.

1. Bring sections down to distilled water and immerse in Solution A for twenty-four hours.

2. Wash with 95% alcohol, then with distilled water.
3. Immerse in Solution B for two and a half hours at 40°C.
4. Wash with distilled water; then reduce in Solution C for five minutes, afterwards washing in tap water.
5. Tone for five minutes in Solution D at 30°C; then wash in tap water.
6. Intensify by immersing in Solution E for five minutes; then rinse in tap water.

   N.B. The above stages must be carried out in the dark-room.
7. Fix in Solution F.

(Note. Fixation should be completed in fifteen to twenty minutes).

8. Wash in tap water; dehydrate; clear and mount.

Results

Tumour cells: reddish to greyish violet. Vascular reticulum; black.

Method 10.23

Bismarck Brown-Methyl Green Method

(for macin, cartilage)

Solution required

A. Bismarck brown 1 % aqueous.

B. Methyl green 0.5% aqueous.

Procedure

Tissues are fixed in Bouin or Zenker and embedded in paraffin wax.

1. Sections are brought down to distilled water; then stained five to ten minutes in Solution A.
2. Wash with 95% alcohol.
3. Stain with Solution B until the preparation appears dark green to the naked eye.
4. Dehydrate with 95% and absolute alcohol; then clear in xylol, and mount.

Results

Cartilage: dark brown. Mucin: light brown. Nuclei of all cells: green.

Notes

The authors state that the method is essentially the same as *List (1885)*.

Method 10.24

Cresylfast Violet-Toluidine Blue-Thionin Method (Ehrlich Method) (for nerve cells)

Solutions required

A.	Cresylfast violet, CNS	2 gm
	Toluidine blue	1 gm
	Thionin (Ehrlich)	0.5 gm
	Ethyl Alcohol 30%	200 ml
B.	Distilled water	200 ml
	Sulphuric or nitric acid, conc.	0.5 ml

Procedure

1. Formalin fixed material is embedded in paraffin wax, and sections 4,c in thickness are fixed to slides with glycerin albumen.
2. Remove wax with xylol.
3. Rinse with absolute alcohol.
4. Pass through 95% alcohol. 5. Pass through 80% alcohol.
6. Immerse slides for five to ten seconds in the staining solution at 80-90°C.
7. Differentiate for one second in solution B.
8. Dip and agitate slides in a beaker of cold distilled water for one second.
9. Differentiate further in 90% and 95% alcohol for one to two seconds in each.
10. Immerse in 80% alcohol for one second.
11. Dip and agitate the slides in the still warm solution A for one to two seconds.
12. Return to 80% alcohol for one second. 13. Repeat steps 11 and 12.
14. Rinse in distilled water.
15. Dehydrate by immersing for one second in each of 80%,. 95% and absolute alcohol.
16. Immerse in xylol for one minute.
17. Immerse in a fresh lot of xylol for three minutes.
18. Mount and examine.

Results

Neurons stand out distinctly against a pale background, and can be followed for a considerable distance. The cytons are stained dark purple emphasizing the blue tint, while the dendrite and axon processes and endings present a somewhat ligbter shade, bluish to reddish. Granules in the cell body as well as in the protoplasm processes appear purple or reddish. Nuclei and nucleoli are well differentiated.

11

HISTOCHEMICAL STAINS AND STAINING PROCEDURE

Histochernical techniques enable the identification and localisation of specific substances within tissues. The methods depend on chemical reactions between the substance to be identified and *localised* in a tissue section, and one or more reagents in which the tissue section is incubated. The histochemist tries to rearrange matters so that the end product of the chemical reaction is both coloured and insoluble, and therefore easily visible on microscopy.

Now the histochemists are equipped with adequate tools for further studies and therefore the 1 number of substances for which histochemical methods have been devised is increasing rapidly. Adequate histochemical methods now exist for the demonstration of *carbohydrates*, *proteins*, *lipids*, *nucleic acids*, *enzymes* etc. At present, in this chapter, we consider the staining techniques for carbohydrates, proteins and lipids. Rest are discussed elsewhere in this book.

CARBOHYDRATES

Although the term "*carbohydrates*" embraces all the sugars and their derivatives, the only such substances available in sections of fixed tissue are those in which the sugars form parts of *macromolecules*. The compounds contain mainly carbon, *hydrogen* and *oxygen*. It is not possible to demonstrate many of the individual carbohydrates at the present time because of their solubility.

Glycogen

This is the only member of the *polysaccharide*. Group which can normally be demonstrated histochemically in tissue sections. Glycogen, which is polymerised glucose, is a normal polysaccharide it is a storage carbohydrate and accumulate in the hepatocytes in the liver. It is also demonstrable in skeletal and cardiac muscle, as well as in many organs or the new born. *Glycogen* will rapidly break down to glucose after death, so it is imperative to obtain rapid and adequate fixation, A great deal has been written about the fixation of glycogen, especially in regard to its possible solubility in water.

Conflicting opinions are still eipressed about the choice of fixative. weather to use alcoholic fixatives, Picric acid fixatives or any standard solution. Lillie showed that little or no loss of glycogen was observed after normal saline fixation, followed by a lengthy wash in running tap water. Personally I prefer to use Gendre's fixative at 4°C on thin slices of tissue. Murcury containing fixatives should be avoided.

Glycogen can be routinely demonstrated by Best's *carmine*, Schiff methods, silver techniques or by iodine. The silver and iodine methods are the least specific and are little used today. The best's carmine method can prove capricious and a known positive control should always be used.

Method 11.1

Best's carmine method

Reagents required

(*a*) Carmine
(*b*) Potassium carbonate
(*c*) Potassium chloride
(*d*) Ammonia (.880)
(*e*) Methyl alcohol
(*f*) Absolute alcohol-
(*g*) Distilled water

Preparation of Solutions

(*a*) Best's carmine stock solution

Carmine	2 g
Potassium carbonate	1 g
Potassium chloride	5 g
Distilled water	60 ml

Boil the solution gently for 5 min, allow to cool, then filter. To the filtrate add 20 ml of ammonia (.880).

(*b*) Best's carmine staining solution

Stock solution	12 ml
Ammonia (.880)	18 ml
Methyl alcohol	18 ml

(*c*) Best's differentiator

Absolute alcohol	8 ml
Methyl alcohol	4 ml
Distilled water	10 nil

Sections

Paraffin sections

Freeze dried sections (recommended)

Frozen sections

Procedure

(*a*) Bring sections from xylene to 70 percent alcohol

(*b*) Place sections in 1 per cent celloidin — 5 min

(*c*) Transfer sections to tap water

(*d*) Stain in alum haematoxylin — 10-15 min

(*e*) Wash briefly in tap water

(*f*) Stain in solution (*b*) (Best's carmine staining solution) — 30 min

(*g*) No water, rinse sections in 2 changes of solution (*c*)

(*h*) Wash sections in 90 per cent alcohol briefly 20.

(*i*) Place in absolute alcohol

(*j*) Transfer to xylene

(*k*) Mount in DPX.

Result

Glycogen : red

Nuclei : blue.

Method 11.2

Baner-Feulgen Method

Reagent required

(*a*) Chromium trioxide

(*b*) Schiff's reagent

(*c*) Sodium metabisulphite

(*d*) Hydrochloric acid

(*e*) Distilled water

Staining Solutions

(*a*)	Oxidising solution (4 per cent chromic acid)	
	Chromium trioxide	4 g
	Distilled water	100 ml
(*b*)	Schiff's reagent	
(*c*)	Sulphurous acid rinse	
	10 per cent sodium metabisulphite	5.0 ml
	Hydrochloric acid 5.0 ml Distilled water	90 ml

Sections

This method is recommended for the demonstration of glycogen in frozen sections (crystal, etc.).

Procedure

(*a*)	Bring all sections to water	
(*b*)	Place in solution (*a*) (chromic acid)	30 min
(*c*)	Wash well in tap water 10 min	
(*d*)	Place sections in Schiff's reagent (solution 2)	15 min
(*e*)	Transfer to solutions (*c*) (sulphurous acid rinse)	2 min
(*f*)	Transfer to fresh solution (*c*) (sulphurous acid rinse)	2 min
(*g*)	Transfer to fresh solution (*c*) (sulphurous acid rinse)	2 min
(*h*)	Wash in tap water	5 min
(*i*)	Counterstain in haematoxylin	2-5 min
(*j*)	Wash in tap water	5 min
(*k*)	Dehydrate through graded alcohols to xylene	
(*l*)	Mount in DPX	

Results

Glycogen : red

Nuclei : blue

Remarks

This technique gives the best result for demonstrating glycogen in frozen sections other carbohydrate substances are over oxidised and are not stained.

Method 11.3

Preparation of Schiff's reagents

(*a*) ***de Tomasi (1936)***

Dissolve I g of basic fuchsin in 200 ml of distilled water -and boil. Shake the solution for 5 min and allow to cool. When the temperature is down to 50°C filter and to the filtrate -add 20 ml N-hydrochloric acid. Cool to 25°C and add 1 g of sodium metabisulphite. Store the solution in the dark, overnight. To this solution, add 2 g of activated charcoal and shake for 1 min. Filter and store the filtrate in a dark bottle at 4°C.

Note. Always allow the aliquot of solution to reach room temperature before use and discord it after use to avoid contamination of stock solution.

(*b*) ***Barger and Oe Lamater (1948)***

Dissolve 1 g of basic fuchsin in 400 ml of boiling distilled water, cool to 50°C and filter. To the filtrate add 1 ml of thioryl chloride. Stand in the dark overnight. Add 2 g of.activated charcoal, shake well and filter. Store the filtrate at 4°C in a dark bottle.

Note. Allow the aliquat of Schiff's solution to reach room temperature before use and discard it after use.

Method 11.4

PAS Reaction Reagents Required

(*a*) Periodic acid
(*b*) Basic fuchsim
(*c*) Hydrochloric acid
(*d*) Sodium metabisulphite
(*e*) Activated charcoal
(*f*) Distilled water

Solution 1.

Periodic acid 1 per cent

Periodic acid	1 g
Distilled water	100 ml

Solution 2

Schiff's reagent (de Tomasi)

Sections

All types.

Procedure

(*a*) Bring all sections to water
(*b*) Place sections in (periodic acid) solution (*a*) 5-8 min

(*c*) Wash in tap water	3 min
(*d*) Wash in distilled water	1 min
(*e*) Treat with Schiff's reagent (solution 2)	15 min
(*f*) Wash in tap water	10 min
(*g*) Counterstain in haematoxylin	
(*h*) Wash in tap water	5 min
(*i*) Differentiate if necessary in I per cent acid alcohol	
(*j*) Wash in tap water	5 min
(*k*) Dehydrate through graded alcohols to xylene	
(*l*) Mount in DPX	

Result

PAS-Positive material : magenta Nuclei : blue.

Method 11.5

Acid Mucosubstances : Alcian Blue modified

Reagents required

(*a*) Alcian Blue
(*b*) 3 per cent acetic acid
(*c*) 10 per cent sulphuric acid
(*d*) Mayer's carmalum

Staining Solutions

Alcian Blue pH 2.5

Alcian Blue	1 g
3 per cent acetic acid	100 ml

Alcian Blue pH 0.2

Alcian Blue	1 g
10 per cent sulphuric acid	100 ml

Sections

All types, freeze dried recommended.

Procedure

(*a*) Bring all sections to water	
(*b*) Stain in Alcian blue solution of choice	5 min
(*c*) Wash briefly in distilled water	
(*d*) Counterstain in Mayer's carmalum	2 min
(*e*) Wash in tap water	
(*f*) Dehydrate through graded alcohols to xylene	

(*g*) Mount in DPX

Results

Alcian Blue pH 0.2 : strongly sulphated acid mucosubstances blue.

Alcian Blue pH 2.5 : most acid mucosubstances blue.

Remarks

(*a*) Alcian Blue stains should be filtered before use

(*b*) Counterstaining should always be light

(*c*) For critical staining see.

Method 11.6

Alcian Blue-CEC Method

Reagent required

(*a*) Alcian Blue

(*b*) Acetate buffer pH 5.8

(*c*) Magnesium chloride

Stock Alcian Blue stain

Alcian Blue	50 mg
Acetate buffer pH 5.8	100 ml

This stock solution is used with the necessary amount of magnesium chloride. To obtain the following molarities the stated figure Alcian Blue solution.

0.6 M=1.2 g, 0.3 M = 6.1 g, 0.5 M=10.15 g, 0.7 M = 14.2 g, 0.9 M = 18.3 g.

Sections

All types.

Procedure

(*a*) Bring all sections to water

(*b*) Stain in Alcian Blue solution for at least 4 hours, overnight preferably.

(*c*) Rinse in distilled water

(*d*) Dehydrate through graded alcohols to xylene

(*e*) Mount in DPX

Results

Positive with 0.06 M carboxyl and sulphated mucosubstances.

Positive with 0.3 M weakly and strongly sulphated mucosubstances.

Positive with 0.5 M strongly sulphated mucosubstances.

Positive with 0.7 M highly sulphated connective tissue mucins.

Positive with 0.9 M keratan sulphate only.

Remarks

(*a*) Background staining can occur with the low molarity solutions.

(*b*) For comments of specificity of method.

(*c*) Counterstain is optional.

Method 11.7

Alcian Blue-PAS Method

Reagents required

(*a*) Alcian Blue solution

(*b*) 1 per cent periodic acid

(*c*) Schiff's reagents

Staining Solution

(*a*) Alcian Blue pH 2.5.

(*b*) Schiff's reagents.

Procedure

(*a*)	Bring all sections to water	
(*b*)	Stain in Alcian Blue solution	5 min
(*c*)	Wash in distilled water	
(*d*)	Oxidise in I per cent periodic acid	5 min
(*e*)	Wash well in distilled water	
(*f*)	Place in Schiff's reagent	8 min
(*g*)	Wash well in running tap water	10 min
(*h*)	Counterstain lightly in Mayer's haematoxylin if required.	
(*i*)	Differentiate in 1 per cent acid alcohol, blue, etc.	
(*j*)	Dehydrate through graded alcohols to xylene.	
(*k*)	Mount in DPX	

Results

Acid mucosubstances : blue

Neutral mucosubstances : red Mixtures : purple

Remarks

(*a*) A good method to separate acid and neutral mucosubstances.

(*b*) As Cook states, the method if negative can be taken to mean that a given substance is unlikely to be a mucosubstances.

Method 11.8

Acid Mucosubstances : Dialysed Iron Method
Modified by Muller; and Mowry,

Reagents required

(*a*) Ferric chloride
(*b*) Acetic acid
(*c*) Hydrochloric acid
(*d*) Potassium ferrocyanide
(*e*) Distilled water.

Preparation of Solutions

(*a*) Stock colloidal iron solution

29 per cent ferric chloride	2.2 ml
Distilled water	125 ml

Pour the ferric chloride into boiling distilled water and stir well. The solution will turn dark red; at this point it is removed from the heat and allowed to cool. It is important,that the solution is boiling when the ferric chloride is added.

(*b*) Staining Solution of Colloidal Iron

Glacial acetic acid	5 ml
Distilled water	15 ml
Stock colloidal iron solution	20 ml

(*c*) Acid Ferrocyanide Mixture

Potassium ferrocyanide	2 g
Hydrochloric acid (conc.)	2 ml
Distilled water	98 ml

The potassium ferrocyanide is dissolved in the distilled water and the hydrochloric acid is added to this solution.

Sections

All types.

Procedure

(*a*) Bring all sections to water,
(*b*) Rinse in 12 per cent acetic acid.
(*c*) Stain in solution
(*b*) collodial iron solution 1 h
(*d*) Rinse in 12 per cent acetic acid,
4 charges, 3 min each.

(*e*) Treat section with acid ferrocy anide, solution (*c*) — 20 min

(*f*) Wash in distilled water

(*g*) Wash in tap water — 5 min

(*h*) Counterstain in Mayer's carmalum — 10 min

(*i*) Wash in tap water

(*j*) Dehydrate through graded alcohols

(*k*) To xylene and mount in DPX

Results

Acid mucosubstances; bright blue

Nuclei : red

Remarks

Some workers prefer to use the Feulgen reaction as a counterstain.

Method 11.9

Acid Mucosubstances: Toluidine Blue Method

Reagents required

(*a*) Toluidine Blue

(*b*) Absolute alcohol

(*c*) Distilled water

Staining Solution

Toluidine Blue	100 mg
Absolute alcohol	30 ml
Distilled water	70 ml

PROTEINS

Proteins are the third of the components that go to make up cell cytoplasm, the other two being lipids and carbohydrates. Proteins are to be found in all cells and tissues. The actual demonstration of protein material is not difficult, but the identification of any particular protein is difficult. The simpleproteins are the naturally occurring proteins that upon hydrolysis will yield amino acids only.

The conjugated proteins consist of simple proteins combined with a non-protein material, for example, lipids in lipoproteins and nucleic acids in nucleoproteins. The methods discussed below do not demonstrate the protein molecule as a whole, but are dependent on the presence of certain types of reactive groups or special types of linkages within the protein molecule.

Histo-chemical reactions are available for some, but not all, of the reactive groups of the protein molecules. Some of the important methods are discussed as follows.

Method 11.10

The Ninhydrin-Schiff method

Reagents required

(*a*) Ninhydrin
(*b*) Absolute alcohol
(*c*) Schiff's reagent

Preparation of solutions

(*a*) Ninhydrin 0.5 per cent solution

Ninhydrin	500 mg
Absolute alcohol	100 ml

(*b*) Schiff's reagent

Sections

Freeze dried
Paraffin sections
Cryostat unfixed
Cryostat prefixed

Procedure

(*a*) Bring sections down to 70 per cent alcohol
(*b*) Treat with solution (*a*), ninhydrin, at 37°C overnight
(*c*) Wash in running tap water
(*d*) Place sections in Schiff's reagent, solution (*b*), for 45 min
(*e*) Wash well in running tap water
(*f*) Counterstain if required in haematoxylin
(*g*) Wash in tap water
(*h*) Dehydrate through graded alcohols to xylenes and mount.

Results

α-Amino groups: pink to red

Remarks

(*a*) 1 per cent Alloxan, also in absolute alcohol, may be used instead of 0-5 per cent ninhydrin.
(*b*) Control sections may be needed for other PAS-positive material.

Method 11.11

DNFB method for tyrosine, SH and NH, groups

Reagents required

(*a*) 2-4 Dinitrofiuorobenzene
(*b*) Ethyl alcohol
(*c*) Sodium hydrogen carbonate
(*d*) Sodium hydrosulphite
(*e*) Sodium nitrite
(*f*) Hydrochloric acid
(*g*) H-acid (8 amino- l-naphthol 3, 6-disulphonic acid)
(*h*) Veronal acetate buffer pH 9.4

Preparation of Solutions

(*a*) *Dinitrofluorobenzene solution*

2-4 dinitrofluorobenzene saturated in 90 per cent ethyl alcohol saturated with sodium hydrogen carbonate.

(*b*) *Nitrous acid*

5 per cent sodium nitrite	10 ml
2w hydrochloric acid	40 ml

(*c*) *'H'-acid*

H-acid saturated in 0.1 m veronal acetate buffer pH 9.4.

Procedure

(*a*) Bring sections to absolute alcohol and allow to dry in air
(*b*) Place in DNFB solution 2-16 h (overnight is usually ideal)
(*c*) Rinse in 90 per cent alcohol (3 changes) then tap water
(*d*) Treat with 5 per cent sodium hydrosulphite at 45°C for 30 min
(*e*) Wash in distilled water
(*f*) Treat sections with nitrous acid solution at 4°C for 30 min
(*g*) Wash in cold distilled water
(*h*) Treat with H-acid solution at 4°C for 15 in
(*i*) Wash in tap water
(*j*) Dehydrate through graded alcohols to xylene
(*k*) Mount in DPX

Results

A positive reaction is reddish purple

Method 11.12

Sakaguchi method
modified by Baker

Reagents required

(*a*) Sodium hypocblorite
(*b*) α-Napbthol
(*c*) Sodium hydroxide
(*d*) Pyridine
(*e*) Chloroform
(*f*) 70 per cent Alcohol

Preparation of Solutions

Solution (*a*)

1 per cent Sodium hydroxide

Solution (*b*)

α-Naphthol	1 g
70 percent Alcohol	100 ml

Solution (*c*)

Milton	1 ml
Distilled water	99 ml

Incubating solution (*d*)

Solution (*a*)	2 ml
Solution (*b*)	2 drops
Solution (*c*)	4 drops

Pyridine-chloroform solution (*d*)

Pyridine	30 ml
Chloroform	10 ml

Sections

Freeze dried
Cryostat
Paraffin sections

Suitable Control Sections

Testis

Procedure

(*a*) Bring all sections to water
(*b*) Rinse in 70 per cent alcohol

(*c*) Cover section with incubating solution (*d*), 15 min
(*d*) Drain and blot dry
(*e*) Immerse in pyridine-chloroform, solution (*e*), 2 min
(*f*) Mount in pyridine-chloroform mixture and ring cover slip

Result

Arginine: orange-red

Remarks

The slide should be looked at microscopically immediately.

Method 11.13

DMAB-nitrite method

Reagents required

(*a*) p-Dimethylaminobenzaldehyde
(*b*) Hydrochloric acid
(*c*) Sodium nitrite
(*d*) Acid alcohol (1 per cent)

Preparation of Solutions

(*a*) p-Dimethylaminobenzaldehyde

p-Dimethylaminobenzaldehyde	5 g
Hydrochloric acid (conc.)	100 ml

(*b*) Sodium nitrite

Sodium nitrite	1 g
Hydrochloric acid (conc.)	100 ml

(*c*) Acid alcohol

Hydrochloric acid (conc.)	1 ml
70 per cent Alcohol	99 ml

Sections

Freeze dried
Paraffin sections
Cryosta

Suitable control sections

Pancreas, duodenum, pituitary

Procedure

(*a*) Bring sections to alcohol
(*b*) Celloidinise in 0.5 per cent celloidin
(*c*) Place sections in solution (*a*), DMAB 1 min

(*d*) Transfer sections to solution (*b*), sodium nitrite 1 min
(*e*) Wash carefully in tap water 30 sec
(*f*) Rinse sections in solution (*c*), acid alcohol 15 sec
(*g*) Dehydrate through graded alcohols to xylene and mount.

Results

Tryptophan : deep blue

Remarks

This method gives good localisation.

Method 11.14

Million reaction, Baker

Reagent required

(*a*) Mercuric sulphate
(*b*) Sulphuric acid
(*c*) Sodium nitrite

Preparation of Solutions

Solution (*a*)

Distilled water	90 ml
Sulphuric acid (conc.)	10 ml

To this solution add 10 g mercuric sulphate and beat until dissolved. Cool to room temparature and add 100 ml distilled water.

Solution (*b*)

Sodium nitrite	250 mg
Distilled water	10 ml

Staining solution (*c*)

Solution 1	10 ml
Solution 2	1 ml

Sections

Freeze dried
Paraffin sections
Cryostat sections
Celloidin sections

Suitable controls

Pancreas, duodenum

Procedure

(*a*) Bring all sections to water

(*b*) Place sections in a small beaker, add solution (*c*) and boil gently — 2 min
(*c*) Allow to cool to room temperature
(*d*) Wash sections in distilled water — 2 min
(*e*) Repeat wash in distilled water — 2 min
(*f*) Repeat wash in distilled water — 2 min
(*g*) Dehydrate through graded alcohols to xylene and mount in DPX

Result

Tyrosine: red, pink or yellowish red.

Method 11.15

Diazotisation-coupling method for tyrosine

Reagents required

(*a*) Sodium nitrite
(*b*) Acetic acid
(*c*) 8-Amino-l-naphthol-5-sulphonic acid (S-acid)
(*d*) Potassium hydroxide
(*e*) Ammonium sulphamate
(*f*) Hydrochloric acid

Preparation of solutions

(*a*)	Incubating solutlon	
	Sodium nitrite	6.9 g
	Acetic acid	5.8 ml
	Distilled water	94 ml
(*b*)	8-amino- I -naphthol-5-sulphonic acid	1 g
	Potassium hydroxide	1 g
	Ammonium sulphamate	1 g
	70 per cent alcohol	100 ml

Sections

All types

Suitable control sections

Pancreas

Procedure

(*a*) Place sections in incubating medium at 4°C overnight in the dark

(*b*) Rinse in distilled water at 4°C
(*c*) Treat with solution (*b*) at 4°C for 1 h also in the dark
(*d*) Rinse in 3 changes of 0-1 NHCl 5 min each
(*e*) Wash in tap water 10 min
(*f*) Dehydrate through graded alcohols to xylene
(*g*) Mount in DPX

Results

Tyrosine-containing proteins: purple-red.

Method 11.16

Ninhydrin Meltiod (for Amino Groups)

Solutions required

A. Celloidin 0.5% in equal volumes of absolute alcohol and ether.
B.N inhydrin (triketohydrindene hydrate), 0.5% aqueous.

Procedure

1. Fix pieces of fresh material in 10% formalin for two to four hours.
2. Wash in water and cut frozen sections.
3. Wash in 70% alcohol, followed by 90% and absolute alcohol.
4. Coat sections on slides with the celloidin solution, then allow the solvent to evaporate for a few minutes.
5. Immerse sections in solution B (Ninhydrin) for one minute at 90°C.
6. Wash in water.
7. Mount in Aquamount or glycerine or glycerine jelly.

Results

Alpha-amino acids and proteins: blue to violet.

Notes

(a) According to Serra, the reaction is given not only by all amino acids (except proline and hydroxyproline), peptides and proteins, but by amines, aldehydes, sugars with free aldehyde and keto groups,-and by ammonia and ammonium salts.
(b) Readers are referred to Serra's paper for more detailed information.

LIPIDS

The term "lipid" is applied to a chemically heterogenous group of substances which are extracted from tissues by nonpolar organic solvents such as chloroform and ether. These substances vary greatly in structural complexity but are built from a limited number of simpler molecules joined together in different ways. These components substances do not occur in large amounts in the free state in lining tissue but are present as metasolic precursors of the lipids themselves.

There is no agreement on bow lipids should be classified, it is probably easiest to make the following subdivisions.

(*a*) Simple lipids

(*b*) Compound lipids

(*c*) Derived lipids

Lipids in general terms can be termed as compounds of long chain fatty acids with an alcohol.

Simple lipids are esters of both saturated and unsaturated long chain fatty acids with alcohols compound lipids contain a non-lipid group, as well as long chain fatty acids and an alcohol. On the other hand, derived lipids contain. The fatty acids that are produced by hydrolysis of the simple and compound lipids. The fatty acids can be either saturated or unsaturated.

Some individual methods for staining lipid materials are given below:

Method 11.17

Oil Red O (Lillie and Ashburn)

Reagents required

(*a*) Oil Red O

(*b*) Triethyl phosphate

(*c*) Distilled water

Preparation of staining solution

Oil Red O	1 g
Triethyl phosphate	60 ml
Distilled water	40 ml

The distilled water is added to the triethyl phosphate, the dye is then added, the mixture is Feated to 100°C for 5 min and is stirred constantly. The mixture is filtered when hot and a gain when cool. This solution will keep as a stock solution but must be filtered before use.

Sections

Formalin-fixed frozen sections, free-floating

Cryostat post-fixed, preferably free-floating

Procedure

(*a*) Wash sections in distilled water
(*b*) Place sections in 60 per cent triethyl phosphate
(*c*) Stain sections in Oil Red O solution at 20°C for 15 min,
(*d*) Wash sections in 60 per cent triethyl phosphate for 30 sec
(*e*) Wash sections in distilled water
(*f*) Stain sections in haematoxylin for 1 min
(*g*) Wash sections in tap water for 5 In
(*h*) Mount in glycerin jelly

Results

Lipid material: red Nuclei: blue

Remarks

This method can be employed at 37°C or 60°C if required.

Method 11.18

Sudan Black B

Reagents required

(*a*) Sudan Black B
(*b*) Triethyl phosphate
(*c*) Distilled water

Preparation of staining solution

Sudan Black B	1 g
Triethyl phosphate	60 ml
Distilled water	40 ml

The distilled water is added to the triethyl phosphate, the stain is added to this solution and the mixture is heated to 100°C for 5 min and stirred constantly. The mixture is filtered when hot and once again immediately before use. The solution will keep well as a stock solution, but must be filtered each time it is used.

Sections

Formalin-fixed frozen sections, free-floating Cryostat post-fixed, preferably free-floating

Procedure

(*a*) Wash sections in distilled water
(*b*) Place sections in 60 per cent triethyl phosphate
(*c*) Stain sections in Sudan Black B solution at 20°C for 10 min
(*d*) Place sections in 60 per cent triethyl phosphate for 30 sec

(*e*) Wash in distilled water

(*f*) Stain in Mayer's Carmalum for 3 min

(*g*) Wash in distilled water (*h*) Mount in glycerin jelly

Results

Lipid material, including phospholipids; black

Nuclei: red

Remarks

This technique using triethyl phosphate can be carried out 37°C or 60°C as required.

Method 11.19

Acidic lipids: Nile Blue

Reagents required

(*a*) Nile Blue

(*b*) Acetic acid

(*c*) Distilled water

Preparation of solution

(*a*) Nile Blue 1 per cent

Nile Blue	500 mg
Distilled water	50 ml

(*b*) Nile Blue 0.02 per cent

Nile Blue	10 mg
Distilled water	50 ml

(*c*) Dfferentfator

Acetic acid, conc.	0.5 ml
Distilled	50 ml

Sections

Formalin-fixed frozen sections

Cryostat prefixed

Cryostat post-fixed

Procedure

It is necessary to stain one section by the Oil Red O or Sudan Black methods. This slide is numbered 3. Two sections are required for the following technique:

(*a*) Bring both sections to water

(*b*) Stain sections 1 and 2 in 1 per cent Nile Blue for 5 m in at 60°C

(*c*) Differentiate at 60°C in solution (*c*) for 30 s
(*d*) Wash in tap water
(*e*) Mount section 1 in glycerin jelly
(*f*) Place section 2 in solution (*b*) (0.02 per cent Nile Blue) for 5 min at 60°C
(*g*) Wash in tap water
(*h*) Differentiate section 2 in solution (*c*) at 60°C for 30 s (*i*) Wash in tap water
(*j*) Mount in glycerin jelly

Results

Section 1. Any blue staining that can be compared with a< positive section 3 is taken to be acidic lipid material.

Section 2. Any red staining that can be compare with a. positive result in section 3 is taken to be a non-acidic lipid.

Section 3. Control for above.

Remarks

This methodhas caused much controversy in the literature,. and its specificity is doubtful, it is however well worth applying, when trying to identify an unknown lipid.

Method 11.20

Acid Haematein method

Reagents required

(*a*) Formalin
(*b*) Calcium chloride
(*c*) Potassium dichromate
(*d*) Haematoxylin
(*e*) Sodium iodate
(*f*) Glacial acetic acid
(*g*) Sodium tetraborate (borax)
(*h*) Potassium ferricyanide
(*i*) Distilled water

Preparation of solutions

(*a*) Fixative

Formalin	10 ml
Calcium chloride (anhydrous)	1 g
Distilled water	90 ml

(*b*) Post-chroming solution

Potassium dichromate	5 g
Calcium chloride	1 g
Distilled water	100 ml

(*c*) Acid Haematein solution

Haematein	50 mg
1 per cent Sodium iodate	1 ml
Distilled water	49 ml

Heat the solution to boiling point, allow to cool and add 1 ml of glacial acetic acid.

(*d*) Differentiator

Potassium ferricyanide	250 mg
Sodium tetraborate (borax)	250 mg
Distilled water	100 ml

Sections

For this method unfixed pieces of tissue are recommended.

Procedure

1. Place blocks in fixative, solution (*a*) at 22°C, for 6-12 hrs.
2. Transfer tissue (no washing) to post-chroming solution (*b*) at 22°C for 18 hours.
3. Transfer tissue to fresh post-chroming solution (*b*) at 60°C, for 24 hours.
4. Wash in running tap water for 6 hours.
5. Cut frozen sections 10 μm thrick.
6. Place sections in post-chroming solution (*b*) at 37°C, for 1 hours.
7. Wash well in distilled water for 5 minutes.
8. Stain in acid haematin, solution (*c*) at 60°C for 5 hours.
9. Rinse well in distilled water.
10. Transfer sections to differentiating solution (*d*) at 37°C for 18 hours.
11. Wash in tap water for 10 minutets.
12. Mount in glycerin jelly.

Results

Phospholipids-dark blue Other material may be blue.

Remarks

This method is only specific when a control is submitted to the pyridine extraction method.

Method 11.21

PAS reaction

Reagents required

(*a*) Periodic acid
(*b*) Schiff's reagent
(*c*) Distilled water

Preparation of solutions

(*a*) Periodic acid

Periodic acid	500 mg
Distilled water	100 ml

(*b*) Schiifs reagent

Sections

Cryostat post-fixed
Cryostat pre-fixed
Formalin-fixed frozen sections
Freeze dried paraffin sections

Procedure

(*a*) Bring sections down to water	
(*b*) Transfer sections to solution (*b*)	5 min
(*c*) Wash in tap water	3 min
(*d*) Place in Schiff's reagent	20 min
(*e*) Wash in tap water	20 min
(*f*) Counterstain in haemalum	5 min
(*g*) Wash in tap water	
(*h*) Differentiate in 1 per cent acid alcohol	5 s
(*i*) Wash in tap water	
(*j*) Mount in glycerin jelly	

Results

Glycolipids, mucins, etc.: red Nuclei: blue

Remarks

It is necessary to use the following controls when applying this method for glycolipids.

(*a*) Oil Red O or Sudan Black to confirm site of possible glycolipid.
(*b*) Aldehydes to be blocked.

Note. If formalin-fixed frozen sections are used, they must be well washed in several changes of distilled water remove any free aldehydes. If cryostat sections are used, alcohol should be used as a fixative.

Method 11.22

Perchloric Acid -Naphthoquinone reaction

Reagents required

(*a*) 1 : 2-Naphthoquinone-sulphonic acid
(*b*) Ethanol
(*c*) Perchloric acid
(*d*) Formaldehyde
(*e*) Distilled water

Preparation of solution

1 : 2-Naphthoquinone-4-sulphonic acid	12 mg
Ethanol	6 ml
60 per cent Perchloric acid	3 ml
Conc. formaldehyde	0.3 ml
Distilled water	2.7 ml

The ethanol-perchloric acid-formaldehyde-water solution is prepared first and the reagent dissolved in it.

Sections

Formol saline-fixed, frozen sections, free-floating
Formol calcium-fixed, frozen sections, free-floating

Suitable sections for controls

Adrenal gland

Procedure

(*a*) Cut frozen sections and float into formalin. Leove for 7 days
(*b*) Mount sections on slides and dry at room temperature
(*c*) Soak sections in reagent
(*d*) Heat sections in reagent to 60-70°C for 10 min
(*e*) Mount section in 60 per cent perchlorie acid

Results

Cholesterol and its esters : dark blue

Remarks

(*a*) The dark blue colour is stable for a few hours.
(*b*) During the beating, the section should change colour from red to dark blue.

Method 11.23

Digitonin method

Reagents required

(*a*) Ethyl alcohol
(*b*) Digitonin
(*c*) Distilled water

Preparation of sections

(*a*) Ethyl alcohol 50 per cent

Ethyl alcohol	100 ml
Distilled water	100 ml

(*b*) Digitonin solution

Digitonin	500 mg
Solution (*a*)	100 ml

Sections

Formalin-fixed frozen sections free-floating. A control. section is stained by Oil Red O method.

Procedure

(*a*) Incubate sections for 3 hours in solution (*b*) at room temperature
(*b*) Rinse sections in solution (*a*)
(*c*) Float sections onto slides
(*d*) Mount in glycerin jelly

Results

Digitonin section : free cholesterol, birefringent.

Oil Red O- section : free cholesterol, birefringent. Choleste rol ester, stained by Oil Red O.

Method 11.24

Copper-Rubeanic Acid method

Reagents required

(*a*) Copper acetate
(*b*) Rubeanic acid
(*c*) Ethanol
(*d*) Distilled water
(*e*) Ethylenediamine tetra-acetic acid (Disodium), EDTA

Preparation of solutions

(*a*) 0.005 per cent Copper acetate

Copper acetate	5 mg
Distilled water	100 ml

(*b*) 0.1 per cent EDTA

Ethylenediamine tetra-acetic acid	50 mg
Distilled water	50 ml

(*c*) 0.1 per cent Rubeanic acid

Rubeanic acid	50 mg
Absolute alcohol	35 ml
Distilled water	15 ml

Dissolve the rubeanic acid in the absolute alcohol by warming slightly then add the distilled water.

Sections

Cryostat unfixed

Cryostat pre-fixed

Formalin-fixed frozen sections

Procedure

1. Place sections copper acetate solution for 3-5 hours.
2. Wash sections in EDTA solution for 10 seconds.
3. Wash sections again in EDTA solution for 10 seconds.
4. Wash sections in distilled water for 10 minutes.
5. Immerse sections in rubeanic acid solution for 30 minutes.
6. Wash sections in 70% alcohol for 3 mts.
7. Wash sections in running tap water.
8. Mount sections in glycerin jelly or dehydrate through graded alcohol and mount in DPX.

Result

Fatty acid-greenish black.

Method 11.25

Luxol Fast Blue method (*For gross brain sections*)

Solutions required

A.	Luxol fast blue	1 gm
	Alcohol, 95%	1 litre
	Acetic acid, 5% aqueous	5 ml
B.	Lithium carbonate, saturated aqueous	
C.	Solution B	10 ml

Distilled water	1 litre

Procedure

1. Fix brains in 10% formalin, preferably by perfusion, and sliced to the desired thickness as soon as they are firm enough.
2. Store the sliced brain in the fixative, for a minimum time of two weeks for human material, or in the case of laboratory animals, for at least one week.
3. Wash the slices in water for at least six hours.
4. Dehydrate for one hour in each of two changes of 95% alcohol.
5. Stain in solution A for sixteen to eighteen hours at 45-55°C.
6. Remove excess stain by immersing the preparations in 95% alcohol.
7. Rinse in distilled water.
8. Differentiate in several changes of solution C.
9. Refine the differentiation in a number of changes of 70% alcohol.
10. Rinse in distilled water.
11. Store in 10% formalin.

Results

Tracts of white matter stain brilliant blue, contrasting strongly with cellular areas of grey matter, stained very pale green.

Notes

The author states that absence of myelin is evident in pathological demyelinated regions of adult brain and in the non-myelinated areas of very young animals.

Method 11.26

Sudan Black Method

Solutions required

A. Formaldehyde-saline

Formalin (Formaldehyde 40%)	10 ml
Sodium chloride 10% aqueous	7 ml
Distilled water	83 ml

Note. Keep a few pieces of marble chips in the solution to maintain neutrality.

B. Formalin (Formaldehyde 40%) neutral.

Note. Keep a few pieces of marble chips in the bottle.

C.	Potassium dichromate 2.5% aqueous	88 ml
	Sodium chloride 10% aqueous	7 ml

Note. Keep a few pieces of marble chips in the bottle.

D.	Dichromate formaldehyde	
	Solution B	1 volume
	Solution C	19 volumes
E.	Potasium dichromate 5% aqueous	
F.	Gelatine/or embedding	
	Gelatine powder	25 gm
	Water	100 ml
	Sodium-p-hydroxybenzoate	0.2 gm

Sprinkle the gelatine on to the water and leave it to soak for an hour, afterwards warming in an incumbator maintained at 37°C, until all the gelatine has dissolved, then strain through muslin while still warm.

Note..If Sodium p-hydroxybenzoate, which is added to prevent the growth of moulds and bacteria, is not available in the laboratory, then 0.25 to 0'5 gm of Thymol should be used instead.

G.	Formalum (for hardening gelatine)	
	Formalin (Formaldehyde 40%)	20 ml
	Potassium alum 5% aqueous	80 ml
	Keep marble chips in the bottle.	

Note : Both gelatine blocks and gelatine sections may be preserved indefinitely in Formalum, which makes the gelatine very hard, thereby facilitating the cutting of thin sections which are non-sticky.

Important. Formalum must not be used in the acid haematein test for phospholipids as the alum would react with the haematein.

H.	Sudan Black	0.5 gm
	Alcohol 70%	100 ml

Boil for ten minutes under a reflux condenser; then cool and filter.

I. *Carmalum (Mayer).*

Procedure

1. Fix a piece of tissue not more than 3 mm thick in the formaldehyde-saline for an hour.
2. Transfer, without washing, to the dichromate formaldehyde (Solution D) and leave for five hours.
3. Transfer, without washing, to 5% aqueous potassium dichromate and leave for about eighteen hours.

4. Leaving the tissue in the same solution, transfer to the paraffin oven at 60°C for twenty-four hours.
5. Wash in running water for six hours.
6. Leave overnight in the melted gelatine in the oven at 37°C.
7. Cool the gelatine, preferably in a refrigerator.
8. Cutout a rectangular block containing the specimen
9. Immerse the block overnight (or any conveniently longer time) in formalum, placing a marble chip in the capsule or tube.
10. Cut sections 8 to 10μ on the freezing microtome.
11. Transfer a section to 70% alcohol.

    Note. It is best to transfer sections from fluid to fluid, up to stage 16 in a Royal Worcester Porcelain thimble No. a-4756, size 2.
12. Transfer to the Sudan black solution, and leave for ½-4 minutes. (The best period is usually about 2½ minutes.)
13. Wash in 70% alcohol for five seconds.
14. Wash in 50% alcohol for one minute.
15. Wash in water, sinking the section gently with a camel hair brush if it floats.
16. Transfer to Carmalum for two to three minutes. (The optimum time is usually three minutes.)
17. Rinse in distilled water.
18. Transfer the section to a fairly large dish, or a tongue jar of tap water, and leave for two minutes, or any conveniently longer time.
19. Wash again in another large bowl of water.
20. Transfer to a petri dish of water.
21. Float the section on to a slide.
22. Blot away excess water but do not allow the section to dry.
23. Mount in Farrants' medium, or in Aquamount.
24. Attach a cap to hold the coverslip to the slide: then leave overnight in the oven to harden the mounting media, before examining the preparation under the oil immersion objective.

    Note. The slide may be examined after a quarter of an hour, if desired; then returned to the oven to complete the hardening.

Results

Lipids, dark blue or blue-back. Cytoplasm: colourless or pale grey-blue. Chromatin: pink or red.

Note. If the results are not good, another section should be tried with variations of the staining times. Never attempt to judge the colouring until the section is mounted and examined under the oil immersion objective.

It is recommended that the technique be learned on the intestine of the mouse, as it is scarcely possible to fail with this. Cut out a piece of empty intestine about 1 cm long and immerse in *formaldehyde-saline* for five minutes, then open it by a longitudinal cut from one end to the other, taking care not to do any unnecessary damage to the villi. The section should be left only one minute in the Sudan black and two and a half minutes in the *carmalum*.

12

CARBOHYDRATES

The term carbohydrate was originally used to describe compounds containing only carbon, hydrogen and oxygen. Subsequently other substances that are functionally and structurally similar have been included. It is not possible to demonstrate many of the individual carbohydrates at the present time because of their solubility. Various groups of these *carbohydrates* can be shown by the techniques described in this chapter.

The nomenclature of carbohydrates is very confused for technologists and histopathologists and is becoming more so as the result of recent attempts to integrate a more accurate biochemical classification into the loose working classification used by histologists. Traditionally, histological technologists and pathologists have used the terms '*mucins*' and '*acid mucopolysaccharides*' to describe some of the groups of carbohydrates.

Histochemical methods produce little actual chemical information in regard to content of the carbohydrate in comparison to a biochemical test. The term *mucosubstance* was recommended by Spicer, Leppi and Stoward as a suitable name for *carbohydrate-rich*, but biochemically undefined, substances seen in tissues. A histochemical classification of mucosubstances cannot be compared to a biochemical classification because of the lack of specific information forthcoming from the histochemical reaction.

At the same time some of the mucosubstances seen histochemically have not been characterised *biochemically*. Jeanloz recommended that the prefix '*muco*' should be omitted from descriptions of carbohydrates, and Pearse produced a biochemical classification of common mucosubstances omitting the use of the prefix '*muco*', as well as producing a histochemical classification. It would be too confusing in this chapter to relate the mucosubstances demonstrated histochemically to their counterparts in a biochemical classification. The classification shown in Table elsewhere in this chapter goes some way as an introduction in relating the two. It is based upon the writings of Meyer; Spicer, Leppi and Stoward; Pearse; and Cook .

The term mucosubstance will be used throughout this chapter and can be defined as substances which contain hexosamine sugars in combination with differing amounts of proteins, and, in some instances, lipids. Some of these carbohydrates are known as neutral mucosubstances. They usually have free *hexose* groups available and are hence demonstrated by the Periodic Acid-Schiff reaction, ('*periodate reactive*'). Other carbohydrates contain acid *radicals* and are known as acid mucosubstances. The type of acid and its position on the reactive molecule will determine some of the histochemical reactions.

In their paper Spicer, Leppi and Stoward recommended that tissue mucosubstances should be classified by (*a*) site; (*b*) their reactivity to histochemical methods; and (*c*) lability to enzyme digestion. An example of this nomenclature would be '*connective tissue mucosubstance*, *Alcian Blue positive* and resistant to *testicular hyaluronidase*'.

The biochemical terms *homoglycan*, *heteroglycan*, *glycosaminoglycan* and *glycosaminoglucuronoglycan*, are being increasingly used by pathologists and histochemists. These terms, unless used accurately and in their proper context, only lead to further confusion in an already confusing nomenclature. Readers are referred to chapter 10 of Pearse for an explanation of this terminology.

POLYSACCHARIDES

Glycogen

This is the only member of the polysaccharide group which can normally be demonstrated histochemically in tissue sections. Glycogen, which is polymerised glucose, is a neutral polysaccharide; it is a storage carbohydrate and accumulates is the hepatocytes in the liver.

It is also demonstrable in skeletal and cardiac muscle, as well as in many organs of the newborn. Glycogen will rapidly break down to glucose after death, so it is imperative to obtain rapid and adequate fixation. A great deal has been written about the fixation of glycogen, especially in regard to its possible solubility in water. Conflicting opinions are still expressed about the choice of fixative, whether to use

Table 12.1 : Carbohydrates.

(*a*) Polysaccharides-glycogen (periodate reactive)
(*b*) Neutral mucosubstances (periodate reactive)
(*c*) Acid mucosubstances
Sulphated
  (*i*) Connective tissue mucopolysaccharides (periodate unreactive)
    (a) keratan sulphate, heparin and dermatan sulphate (stable to testicular hyaluronidase)
    (b) chondroitin sulphates (labile to testicular hyaluronidase)
  (*ii*) Epithelial sulphomucins (stable to testicular hyaluronidase)
    (a) sulphate esters on vic-glycols
    (b) sulphate esters not on vic-glycols
    (c) a group exist as periodate reactive
Non-sulphated
  (*i*) Connective tissue (periodate unreactive)
    (a) hexuronic acid-rich-hyaluronic acid (labile to testicular hyaluronidase)
    (b) sialic acid-rich
  (*ii*) Epithelial sialomucins (periodate reactivity variable)
    (a) sialic acid-rich (labile to sialidase)
    (b) sialic acid-rich (stable to sialidase)
(*d*) Mucoproteins
(*e*) Mucolipids

alcoholic fixatives, picric acid fixatives or any standard solution. Lillie showed that little or no loss of glycogen was observed after formol saline *fixation*, followed by a lengthy wash in running tap water.

Personally I prefer to use Gendre's fixative at 4°C on thin slices of tissue. Mercury-containing fixatives should be avoided. During room temperature fixation, glycogen becomes insoluble and the classical streaming artefact is produced; this may be avoided by using *Gendre's* fixative at 4°C or by freeze drying.

Leske and Mayersbach and Lake reported that unfixed cryostat sections are superior to conventional paraffin processing for the demonstration of glycogen. *Celloidinisation* is necessary to avoid loss of unfixed *glycogen* during staining.

Glycogen can be routinely demonstrated by Best's Carmine, Schiff methods, silver techniques or by iodine. The *silver* and *iodine* methods are the least specific and are little used today. The Best's *Carmine* method can prove capricious and a known positive control should always be used; accurate differentiation is also an important aspect of the technique.

The PAS methods are probably the most specific available for glycogen, although other reactive *carbohydrate* substances will be positive. For this reason enzyme digestion techniques and positive controls should be employed.

Diastase Digestion of Glycogen

This technique allows for the specific demonstration of glycogen when using the Periodic Acid-Schiff and Best's Carmine methods. Diastase will digest glycogen from a section whilst other PASpositive, carbohydrate material, resistant to diastase digestion, will remain. Amylase (*diastase*) is used as a 0.5-1 per cent solution in phosphate buffer pH 6 or distilled water; this solution needs to be freshly prepared.

A watch should be kept on the stock diastase as this seems to deteriorate within six months or so. Saliva, which contains amylase, will usually work well. Sections for digestion are either placed in 0·5-1 per cent diastase at 37°C for 40 minutes or up to one hour at room temperature. Saliva is used for the same times. Diastase can be used before or after staining with Best's Carmine but only before staining with the PAS reaction.

Celloidinisation of Sections

There is considerable dispute in the literature about the necessity for using celloidin. At one time it was considered that celloidinisation prevented loss of glycogen from the section, when in contact with water. A number of workers now consider that little or no loss of glycogen occurs after fixation during staining; however celloidinisation does prevent diastase from digesting glycogen in control sections. Unfixed cryostat sections to be used for the demonstration of glycogen will need to be coated with 0.25 per cent celloidin.

NEUTRAL MUCOSUBSTANCES

These contain sugars and normally have free hexose groups available, but do not contain any free acid radicles or sulphate esters. These mucosubstances are found in the stomach, intestines and prostate. They show a positive PAS reaction (periodate reactive), and a variable result with Southgate's Mucicarmine.

Acid Mucosubstances

This is the largest of the groups of mucosubstances and can, in the first instance, be divided into sulphated and non-sulphated (carboxylated) mucosubstances.

Sulphated

These mucosubstances can also be conveniently divided into two sub-groups, the connective-tissue sulphated mucosubstances, and the epithelial sulphate *mucosubstances* (*epithelial sulphomucins*). The connective-tissue sulphated mucosubstances contain either sulphated glucuronic acids, such as the chondroitin sulphates and heparan sulphate, or keratan sulphate.

The *chondroitin* sulphate-containing acid mucosubstances are labile to testicular hyaluronidase but the other connective-tissue sulphated mucosubstances are resistant to digestion by this enzyme. The connective-tissue sulphated mucosubstances are found in *skin*, *cartilage*, *heart valves* and *aorta*; they are also known as the strongly acidic and strongly sulphated mucosubstances.

They are usually periodate unreactive, stain with AlcianBlueat pH 0.5 and are metachromatic with Azure 'A' at pH 0.5. The epithelial sulphated mucosubstances (epithelial sulphomucins) also contain sulphate esters, but in different combination to those discussed above. They are found in the submandibular salivary glands, and in duodenal and colonic goblet cells.

They are also known as weakly sulphated mucosubstances, and their staining reactions differ from those of the strongly sulphated mucosubstances in that the optimal pH level with Toluidine Blue is raised from 0.5 to 2.0 and with Alcian Blue from 0.5 to 1.5. They are resistant to testicular hyaluronidase and may be periodate reactive with the exception of an epithelial mucosubstance produced by bronchial serous glands, which is Periodic acid-Schiff positive.

Non-Sulphated (Carboxylated)

This group of mucosubstances may also be conveniently separated into sub-groups, by their site, content, and by their response to enzyme digestion techniques. The classification by site is based on whether the substance is found in the epithelium or connective tissue, and the classification by content is based on whether the acid radical present is sialic acid or a hexuronic acid.

Connective-tissue non-sulphated mucosubstances are rich in hexuronic acid and possibly sialic acid. The hexuronic acid-rich mucosubstances contain large amounts of hyaluronic acid. The reactive components are the carboxyl groups. These substances react with Alcian Blue at pH 2.5 and with *metachromatic* dyes at pH 6. They are periodate unreactive, and are labile to testicular *hyaluronidase*. The existence of a sialic acid-rich connective-tissue mucosubstance is still in doubt; it may be a minor component of cartilage.

Epithelial non-sulphated mucosubstances contain as their acid radical, sialic acid, and are often called epithelial sialomucins. The sialic acid is in combination with the reactive carboxyl groups. They show identical reactions with metachromatic dyes and *Alcian Blue* to the hexuronic acid-rich mucosubstances discussed above, but they are usually periodate reactive (PAS-positive). In addition they are not affected by *testicular hyaluronidase*. Most of these *epithelial sialomucins* are labile to sialidase, but Lev and Spicer described a sialomucin which was resistant to sialidase digestion. Cook described these mucosubstances in human gastric pyloric glands.

MUCOPROTEINS

Mucoproteins are complex substances in which the polysaccharide is chemically combined with a protein. The carbohydrate content of mucoproteins chexosamine) exceeds 4 per cent of the total weight, compared with glycoproteins which have less than 4 per cent. There is no specific method for mucoproteins.

They are positive by the PAS method but are negative with the Toluidine Blue technique. It may be necessary to apply a protein method to see if a positive result is obtained at the same site as the PAS-positive material. If this is the case, it may indicate a mucoprotein. The method used to demonstrate mucoproteins is the PAS method. They are found in the beta cells of the pituitary, in basement membranes and other sites.

MUCOLIPIDS

These are similar to *glycolipids* and are a mixture of polysaccharide and fatty acid complexes. They are found in tissues as cerebrosides and *gangliosides*; they usually give a positive PAS reaction and in frozen sections can be stained by *lipid methods*.

Preparation of Tissue

Sections for the demonstration of carbohydrates can be prepared by the majority of the standard techniques. Freeze drying, followed by vapour fixation, gives the best results. Frozen sections are not ideal because of the deleterious effects of freezing and thawing on localisation.

Paraffin sections normally give acceptable results after either formol saline or formol mercury fixation. There are a considerable number of methods available to demonstrate mucosubstances. Many of these techniques are relatively new and whilst being more complicated the new methods produce more accurate information. The methods listed below and to be found at the end of the chapter are a mixture of the traditional and new methods along with suitable control methods.

Mucicarmine	— Mucosubstances (variable results)	Method 12.3
Periodic Acid-Schiff	— Neutral muco substances; glycogen	Method 12.4
Bauer—Feulgen	— Neutral muco substances; glycogen	Method 12.1
Best's Carmine	— Glycogen	Method 12.2
Alcian Blue (pH 2.5)	— Acid mucosubstances	Method 12.7
Alcian Blue (CEC)	— Sulphated and carboxylated muco substances	Method 12.8
Dialysed Iron	— Acid mucosubstances	Method 12.11
Azure 'A'	— Acid mucosubstances	Method 12.12
Toluidine Blue	— Acid mucosubstances	Method 12.13
Low Iron Diamine	— Acid mucosubstances	Method 12.14
Alcian Blue-PAS	— Neutral mucosubstances and acid muco substances	Method 12.9
Aldehyde Fuchsin-Alcian Blue	— Sulphated and carboxy lated mucosubstances	Method 12.10
High Diamine	— Sulphated and carboxylated mucosubstances	Method 12.15
'Mild' Methylation	— Sulphated muco substances	Method 12.17
Methylation and Saponification	— Carboxylated muco substances	Method 12.18
Sialidase Digestion	— Sialic acid containing carboxylated mucosubstances	Method 12.20
Hyaluronidase Digestion	— Hyaluronic and Chori-droitin sulphate containing muco substances;	Method 12.19
Aldehyde Blocking	— 1,-2 glycol groups	Method 12.16

BAUER-FEULGEN

Bauer suggested this method for the demonstration of glycogen. Oxidation of the glycogen is carried out with 4 per cent chromic acid, followed by staining with Schiff's reagent. Lillie pointed out that carbohydrates other than those of 1,-2 glycol groups will be over-oxidised and hence will not affect the staining. The advantage of this method is that it gives superior results on cryostat or *frozen sections* to the Best's Carmine method.

BEST'S CARMINE

This method was introduced by Best. It has been modified slightly over the years, but basically, the method remains the same. The mechanism of staining is unknown. It is, however, reasonably selective in staining glycogen. Mucin is also stained, but is pale in comparison to the glycogen. If difficulty is encountered, a control section incubated with diastase can be used.

The method occasionally fails to work and a section known to contain glycogen should always be used as a control. The solubility of glycogen in water, according to some workers, necessitates the use of celloidin during the staining procedure; in practice, however, this is not necessary.

The staining method, while ideal for the demonstration of glycogen in paraffin sections and freeze dried material, is not as satisfactory for frozen sections and to demonstrate glycogen in this type of section, it is advisable to use the Bauer—Feulgen method discussed above.

Mucicarmine Method

This technique is an old-established method for acid-mucosubstances. It is empirical and it appears to have relatively high specificity for acid mucosubstances, although the reaction that takes place is not known. Southgate added aluminium hydroxide to the staining solution; this produced more contrast between the mucin and the background as well as making the method more reliable.

PAS Reaction

This technique is the one on which much of the carbohydrate *histochemistry* depends, and is a method in routine use in many laboratories.

The reaction is based upon the fact that periodic acid will oxidise 1,2-glycol groups to produce aldehydes which are coloured by Schiff's reagent. The PAS reaction can be used to demonstrate a wide variety of substances, and because of this, it is necessary to use blocking techniques and controls to produce specific results whenever possible.

Oxidation

Periodic acid is the standard oxidizing agent, although it is possible to use others. The use of chromic acid in the *Bauer-Feulgen* technique has already been described; potassium permanganate and lead tetra-acetate may be used when demonstrating *mucin*. Periodic acid attacks 1,2-glycols and 1,2-amino-alcohols, breaking the carbon chain and oxidizing the adjacent groups of aldehydes. An advantage of using periodic acid

Table 12.2: Oxidizing Agents.

Oxidizing agents	*Strength*	*Solvent*	*For demonstrating*
Periodic acid	0-5%	Distilled water	Carbohydrates
Periodic acid	1-0%	Distilled water	Carbohydrates
Potassium permanganate	5%	Distilled water	
Lead tetra-acetate		Acetic acid	Mucin
Chromic acid	4 %	Distilled water	Glycogen
Periodic acid	0-8%	90% Alcohol	Carbohydrates
Performic acid	Soln.	*See in the text*	Lipids
Peracetic acid	Soln.		Lipids

as the oxidizing agent is that it will rarely over-oxidise the material beyond aldehydes. Oxidation is carried out for five minutes with a 0.5 per cent aqueous solution. This solution is reasonably stable, and may be used repeatedly for at least one month.

Schii 's Reagent

This reagent, developed for chemical work, was introduced by Schiff. There are many modifications of the original formula and all the accepted ones give a good result. It is produced by the reaction of basic fuchsin with *sulphurous acid* to produce a colourless solution, fuchsin-sulphurous acid, or, as it is more commonly known, *Schiff's reagent*. Sulphurous acid breaks the *quinoid structure* of *basic fuchsin*.

In this state, the solution is clear, but if it is then applied to a section containing an aldehyde, the *quinoid structure* is restored and a coloured product is formed at the site of the *aldehydes*. The preparation of Schiff's reagents can cause difficulty. Not all batches of basic fuchsin are the same and some will fail to work for unknown reasons.

The instructions should be followed carefully for the type of Schiff's reagent being prepared. Most techniques call for the addition of acid to the hot distilled water-dye solution. In the author's experience, if the dye solution is too warm, a precipitate is formed, which increases on storage. Schiff's reagent should be discarded after use, as contamination can occur if it is returned to the container. It should be stored at 4°C in a dark bottle, and freshly prepared each month. With these three conditions, few problems should be encountered with the method. Casselman stated that maximum colour intensity is reached after 10 minutes staining time, while other workers leave for 20 minutes or longer.

After treatment with Schiff's reagent, many techniques indicate sulphite rinses before washing in water. [Pearse considered that for use with the PAS reaction in routine histology and histochemistry, this is not necessary.] In practice, very little difference can be detected. Two variations of Schiff's reagent are recommended : (*a*) de Tomasi, *see* Method 12.5; and (*b*) Barger and DeLamater, see Method 12.5.

ALCIAN BLUE

This dye, whilst not being absolutely specific, is considered the most specific dye available for acid mucosubstances. As a technique in histochemistry it was introduced by Steedman. It is a copper phthalocyanin dye giving a brilliant green-blue colour. It is thought that Alcian Blue stains by salt linkage to the acidic groups in acid mucosubstances. If the staining time is kept short, only acid mucosubstances will be coloured.

If the pH of the Alcian Blue is altered it is possible to distinguish between some of the different acid mucosubstances. At pH 2.5 most of the acid mucosubstances are stained. By lowering the pH to 1.0 both the weakly and strongly sulphated acid mucosubstances will be demonstrated, whilst at pH 0.2 only the strongly sulphated mucosubstances will take up the stain. As Cook points out, frequently no clear cut separation of the different acid mucins can be obtained.

An alternative technique is to use the critical electrolyte concentration (CEC) method introduced by Scott and Dorling. The technique is based on the assumption that electrolytes, e.g. magnesium chloride, when incorporated into the Alcian Blue staining solution will compete with Alcian Blue molecules for the reactive constituents of the acid mucosubstances.

Below 0.06 mol/l magnesium chloride, both carboxylated and sulphated mucosubstances will stain; above 0.3 mol/l, only the sulphated mucosubstances will take up the Alcian Blue. The CEC technique has recently been criticised by Goldstein and Horobin on theoretical grounds.

The Alcian Blue method can be combined with the PAS reaction (to distinguish between acid and neutral mucosubstances). It may also be used with aldehyde fuchsin to separate the sulphated and carboxylated mucosubstances.

COLLOIDAL IRON METHOD

The technique to demonstrate acid mucopolysaccharides by using colloidal iron was first described by Hale. This method is not specific, but gives reproducible results which are acceptable when used in conjunction with other methods. The method depends upon the affinity of acidic-tissue components for colloidal iron at low pH levels. The acidic tissue combines with the iron, which is then demonstrated by the Prussian Blue reaction.

It is necessary to stain a control slide by the Prussian Blue method alone, because of the possibility of iron already present in the section. The pH of the colloidal iron solution is important. Mowry (1958) stated that if the colloidal iron solution is below pH 1·3, the technique is more selective. Many workers have produced modifications of the original method to increase its specificity.

The technique given is that published by Mowry , after Muller. The Colloidal Iron method is frequently used as a combination technique, e.g. with the PAS method, or using a neutral red counterstain.

Metachromatic Staining

The majority of acid mucosubstances are capable of exhibiting metachromasia under the correct conditions. A number of dyes, of which Azure 'A', thionin and Toluidine Blue are the most popular, can be used. These dyes, when used in dilute solutions, will react with acid mucosubstances to produce a colour different from that of the dye itself, whilst the remaining tissue components stain the colour of the dye.

The dyes given above are blue but stain acid mucosubstances red. This red staining is termed 'metachromatic' and the blue staining 'orthochromatic'. The metachromatic staining produced with these dyes is to some extent alcohol-labile, and for this reason sections should be mounted in glycerin jelly.

Some workers feel that true metachromasia is only seen after dehydration through graded alcohols, whilst others recommend dehydrating through tertiary butyl alcohol. The less intense metachromatic staining is certainly removed by alcohol treatment. I feel that the sections should be examined in distilled water before mounting as well as after mounting. If the metachromatic dyes are applied to acid mucosubstances at different pH levels, it is also possible to identify separate acid mucosubstances, as the acidic strongly-sulphated mucins are positive at pH 2.0, but negative at pH 6.0. The method 12.12 produces a reasonably alcohol-resistant metachromasia, while Toluidine Blue is a general technique.

Diamine Methods

These were first introduced into histochemistry by Spicer; N. N'dimethyl-phenyl-enediamine-dihydrochloride can be used to demonstrate a number of different mucins. The methods involve the oxidation of a mixture of the meta and para forms of the above salt by ferric chloride to form a black substance which is selectively bound to certain mucosubstances.

When the diamine salt is used in low concentrations (low diamine method), most sialomucins and sulphated mucosubstances stain. At a higher concentration of the diamine salts (high diamine method) only the sulphated mucosubstances are demonstrated. Both methods are followed by staining with Alcian Blue at pH 2.5 to show *nonsulphated acid* mucosubstances.

BLOCKING AND ENZYME METHODS

These techniques are employed in histochemical techniques for mucosubstances to improve the specificity of the method or to realise more information. These control techniques are divided into two groups

Blocking

This is a method in which a '*blocking*' agent will combine with a tissue element chemically to prevent a colour reaction taking place with a dye solution. In this situation the presence of a reacting radical can be shown by its failure to stain in one section, whilst staining in a parallel section without the blocking agent. These blocking techniques can be either '*irreversible*' or '*reversible*'.

The first type of reaction will permanently destroy the reactive groups within the section. The second will temporarily block the reactive groups by formation of chemical derivatives; this type of effect can be reversed by using suitable chemical solutions. An example of blocking techniques is the *aldehyde blocking* in the *Periodic Acid-Schiff reaction.*

Enzyme Digestion Methods

These techniques are used where it is known that an enzyme will digest specific acid mucosubstances. Two sections are stained, one after the enzyme treatment and one with no treatment. This should show the specific substance by its disappearance from the treated section. Neither the blocking nor the digestion techniques are infallible; with the enzyme methods failure is often due to the use of impure enzymes and for this reason suitable positive and negative controls should be used.

Use of Controls

The necessity of using controls with blocking and enzyme methods is perhaps obvious, but it is a point that is frequently overlooked. Cook discussed the matter of suitable controls in detail; he advocated the use of positive and negative controls along with the test section. Negative controls are required in the case of non-specific reactivity.

To prevent this, along with the test section, a duplicate test section and a second positive control should be employed. These two sections will pass through all the reagents except the actual blocking agent or enzyme, including any buffers and solvents used.

Positive controls are sections of tissue known to contain the mucosubstance that is to be demonstrated. They are treated in a similar manner to the test section. In all, Cook recommends that as well as the '*test section*' the three controls discussed above should be used.

BLOCKING TECHNIQUES

Blocking of Aldehydes

This can be carried out by using a number of different methods. Treatment with acetic anhydride will block 1:2 glycol groups by acetylation. This technique is reversible by treatment with potassium hydroxide (saponification). Method 12.16 uses three sections. A negative result with the test section with this method indicates that 1:2 glycol groups were originally responsible for the colouration of Schiff's reagent.

Methylation

This technique can be used to block one or more of the reactive groups of acid mucosubstances. The method uses a methanolhydrochloric acid solution. Spicer divided the method into 'mild' when used at 37°C for 4 hours, and '*active*' when employed at 60°C for 4 hours. Using mild *methylation*, only *carboxyl groups* are blocked, by forming methyl esters.

In using '*active*' methylation at 60°C, all carboxyl and sulphate esters are blocked; the carboxyl groups are methylated whilst the sulphate groups are hydrolysed. This situation shows both reversible blocking, for the treatment with a strong alkali will restore the staining characteristics of the carboxyl groups, but the *sulphate groups* being hydrolysed are irreversibly blocked. The methylation methods given as methods 32 and 33 are those from Spicer.

Saponification

This is the restoration of the reactive carboxyl groups after *methylation*. 1 per cent *potassium hydroxide* in 70 per cent alcohol is usually used. The method is of doubtful value since it is known that false positive staining and false negativity are frequent after *saponification*.

Sulphation

This is the technique of introducing sulphate esters into mucosubstances to enable a subsequent reaction with metachromatic dyes. Lewis and Grillo described a method for sulphating glycogen.

ENZYME DIGESTION METHODS

Three enzymes are commonly used in the elucidation of acid mucosubstances. The first, diastase, has been considered in the discussion of glycogen. The other enzymes are hyaluronidase and sialidase.

Hyaluronidase

This enzyme will digest hyaluronic acid and the chondroitin sulphates A and C. For routine purposes testicular hyaluronidase is used for the digestion techniques. It is normally used as a 0.1 per cent solution in phosphate buffer.

Sialidase

Sialidase (neuraminidase) is used to digest the majority of sialic acid-containing mucosubstances. Tissue sections are incubated for 18 hours in the enzyme solution. Calcium chloride is included as an activator for the enzyme. Following treatment the sections are stained by Alcian Blue (pH 2.5). As indicated earlier some sialomucins are resistant to this enzyme digestion; however sulphuric acid hydrolysis will block both forms of sialic acids.

Method 12.1

Glycogen: Bauer-Feulgen Method

Reagents required

(*a*) Chromium trioxide
(*b*) Schiff's reagent
(*c*) Sodium metabisulphite
(*d*) Hydrochloric acid
(*e*) Distilled water

Staining solutions

(*a*) Oxidizing solution (4 per cent chromic acid)

Chromium trioxide	4 g
Distilled water	100 ml

(*b*) Schiff's reagent

(*c*) Sulphurous acid rinse

10 per cent sodium metabisulphite	5.0 ml
Hydrochloric acid	5.0 ml
Distilled water	90 ml

Sections

This method is recommended for the demonstration of glycogen in frozen sections (cryostat, etc.).

Staining method

(*a*)	Bring all sections to water	
(*b*)	Place in solution (*a*) (chromic acid)	30 min
(*c*)	Wash well in tap water	10 min
(*d*)	Place sections in Schiff's reagent (solution *b*)	15 min
(*e*)	Transfer to solution (*c*) (sulphurous acid rinse)	2 min
(*f*)	Transfer to fresh solution (*c*) (sulphurous acid rinse) 2 min	
(*g*)	Transfer to fresh solution (*c*) (sulphurous acid rinse) 2 min	
(*h*)	Wash in tap water	5 min
(*i*)	Counterstain in haematoxylin	2-5 min
(*j*)	Wash in tap water	5 min
(*k*)	Dehydrate through graded alcohols to xylene	
(*l*)	Mount in DPX	

Results

Glycogen: red Nuclei: blue

Remarks

This technique gives the best result for demonstrating glycogen in frozen sections. Other carbohydrate substances are over-oxidised and are not stained.

Method 12.2

Glycogen: Best's Carmine method

Reagents required

(*a*) Carmine
(*b*) Potassium carbonate
(*c*) Potassium chloride
(*d*) Ammonia (.880)
(*e*) Methyl alcohol
(*f*) Absolute alcohol
(*g*) Distilled water

Preparation of solutions

(*a*) Best's carmine stock solution

Carmine	2 g
Potassium carbonate	1 g
Potassium chloride	5 g
Distilled water	60 ml

Boil the solution gently for 5 min, allow to cool, then filter. To the filtrate add 20 ml of ammonia (-880).

(*b*) Best's carmine staining solution

Stock solution	12 ml
Ammonia (.880)	18 ml
Methyl alcohol	18 ml

(*c*) Best's differentiator

Absolute alcohol	8 ml
Methyl alcohol	4 ml
Distilled water	10 ml

Sections

Paraffin sections

Freeze dried sections (recommended)

Frozen sections

Staining method

(*a*) Bring sections from xylene to 70 per cent alcohol

(*b*) Place sections in 1 per cent celloidin — 5 min

(*c*) Transfer sections to tap water

(*d*) Stain in alum haematoxylin- — 10-15 min

(*e*) Wash briefly in tap water

(*f*) Stain in solution (*b*)
(Best's carmine staining solution) — 30 min

(*g*) No water, rinse sections in 2 changes of solution (*c*) 20 s

(*h*) Wash sections in 90 per cent alcohol briefly

(*i*) Place in absolute alcohol

(*j*) Transfer to xylene

(*k*) Mount in DPX

Results

Glycogen: red

Nuclei : blue.

Method 12.3

Mucosubstances: Mucicarmine

Reagents required

(*a*) Carmine
(*b*) Aluminium hydroxide
(*c*) Aluminium chloride (anhydrous)
(*d*) Absolute alcohol
(*e*) Distilled water

Staining solution

Carmine	1 g
Aluminium hydroxide	1 g
Distilled water	50 ml
Absolute alcohol	50 ml Shake, then add
Aluminium chloride	500 mg

The solution is boiled for 3 min. It is advisable to use a 250 ml flask. After boiling, the solution is allowed to cool to room temperature and is made up to the original volume with 50 per cent alcohol and filtered. This solution is stable for about 12 months. For use, dilute 1 : 4 with distilled water. Store at 4°C.

Sections

All types

Staining method

(*a*) Bring sections to tap water	
(*b*) Stain nuclei with haematoxylin	10 min
(*c*) Wash in tap water	5 min
(*d*) Differentiate in I per cent acid alcohol	
(*e*) Wash in running tap water	5 min
(*f*) Stain in staining solution	30 min
(*g*) Wash in tap water	2 min
(*h*) Dehydrate through graded alcohols to	
(*i*) Xylene, and mount in DPX	

Results

Mucosubstances : red Nuclei : blue.

Method 12.4

Carbohydrates: PAS reaction

Reagents required

(*a*) Periodic acid

(*b*) Basic fuchsin
(*c*) Hydrochloric acid
(*d*) Sodium metabisulphite
(*e*) Activated charcoal
(*f*) Distilled water

Solution 1-Periodic acid 1 per cent

Periodic acid	1 g
Distilled water	100 ml

Solution 2-Sch j's reagent (de Tomasi) (see text).

Sections

All types

Staining method

(*a*)	Bring all sections to water	
(*b*)	Place sections in (periodic acid) solution (*a*)	5-8 min
(*c*)	Wash in tap water	3 min
(*d*)	Wash in distilled water	1 min
(*e*)	Treat with Schiff's reagent (solution 2)	15 min
(*f*)	Wash in tap water	10 min
(*g*)	Counterstain in haematoxylin	
(*h*)	Wash in tap water	5 min
(*i*)	Differentiate, if necessary in 1 per cent acid alcohol	
(*j*)	Wash in tap water	5 min
(*k*)	Dehydrate through graded alcohols to xylene and	
(*l*)	Mount in DPX	

Results

PAS-positive material: magenta Nuclei: blue.

Method 12.5

Preparation of Schiff's reagents

(*a*) *de Tomasi*

Dissolve 1 g of basic fuchsin in 200 ml of distilled water and boil. Shake the solution for 5 min and allow to cool. When the temperature is down to 50°C, filter and to the filtrate add 20 ml N-hydrochloric acid. Cool to 25°C and add 1 g of sodium metabisulphite. Store the solution in the dark, overnight. To this solution, add 2 g of activated charcoal and shake for 1 min. Filter and store the filtrate in a dark bottle at 4°C.

Note. Always allow the aliquot of solution to reach room temperature before use and discard it after use to avoid contamination of stock solution.

(*b*) *Barger and DeLamater*

Dissolve 1 g of basic fuchsin in 400 ml of boiling distilled water, cool to 50°C and filter. To the filtrate add 1 ml of thionyl chloride. Stand in the dark overnight. Add 2 g of activated charcoal, shake well and filter. Store the filtrate at 4°C in a dark bottle.

Note. Allow the aliquot of Schiff's solution to reach room temperature before use and discard it after use.

Method 12.6

Diastase digestion for glycogen

Reagents required

(*a*) Malt diastase
(*b*) 0.02 M Phosphate buffer, pH 6.0
(*c*) Sodium chloride

Preparation of solution

Malt diastase	500 mg
0.02 M Phosphate buffer, pH 6.0	50 ml
Sodium chloride	400 mg

Staining method

(*a*) Bring all sections down to water
(*b*) Treat sections with diastase solution for 1 h at 22°C or 40 min at 37°C
(*c*) Rinse in water
(*d*) Employ histochemical method

Remarks

(*a*) This digestion should be carried out before celloidinisation of the sections
(*b*) Human saliva may also be used, instead of malt diastase, for the digestion of glycogen.

Method 12.7

Acid mucosubstances : Alcian Blue

Reagents required

(*a*) Alcian Blue
(*b*) 3 per cent acetic acid

(*c*) 10 per cent sulphuric acid
(*d*) Mayer's carmalum

Staining solutions

Alcian Blue pH 2.5

Alcian Blue	1 g
3 per cent acetic acid	100 ml

Alcian Blue pH 0.2

Alcian Blue	1 g
10 per cent sulphuric acid	100 ml

Sections

All types, freeze dried recommended

Staining method

(*a*) Bring all sections to water	
(*b*) Stain in Alcian Blue solution of choice	5 min
(*c*) Wash briefly in distilled water	
(*d*) Counterstain in Mayer's carmalum	2 min
(*e*) Wash in tap water	
(*f*) Dehydrate through graded alcohols to xylene	
(*g*) Mount in DPX	

Results

Alcian Blue pH 0.2: strongly sulphated acid mucosubstances blue
Alcian Blue pH 2·5: most acid mucosubstances blue

Remarks

(*a*) Alcian Blue stains should be filtered before use
(*b*) Counterstaining should always be light
(*c*) For critical staining see Method 12.8.

Method 12.8

Acid mucosubstances: Alcian Blue (CEC Method)

Reagents required

(*a*) Alcian Blue
(*b*) Acetate buffer pH 5-8
(*c*) Magnesium chloride

Stock Alcian Blue stain

Alcian Blue	50 mg
Acetate buffer pH 5-8	100 ml

This stock solution is used with the necessary amount of magnesium chloride. To obtain the following molarities the stated figure of magnesium chloride should be added to 100 ml of the stock Alcian Blue solution.

0.6M = 1.2g, 0.3M = 6.1g, 0.5M = 10.15g, 0.7M = 14.2 g, 0.9 M = 18.3 g.

Sections

All types

Staining method

(*a*) Bring all sections to water
(*b*) Stain in Alcian Blue solution for at least 4 hours, overnight preferably
(*c*) Rinse in distilled water
(*d*) Dehydrate through graded alcohols to xylene
(*e*) Mount in DPX

Results

Positive with 0.06 M carboxyl and sulphated mucosubstances
Positive with 0.3 M weakly and strongly sulphated mucosubstances
Positive with 0.5 M strongly sulphated mucosubstances
Positive with 0.7 M highly sulphated connective tissue mucins
Positive with 0.9 M keratan sulphate only

Remarks

(*a*) Background staining can occur with the low molarity solutions
(*b*) For comments of specificity of method *see text*
(*c*) Counterstain is optional.

Method 12.9

Alcian Blue-PAS method

Reagents required

(*a*) Alcian Blue solution
(*b*) 1 per cent periodic acid
(*c*) Schiff's reagent

Staining solutions

(*a*) Alcian Blue pH 2.5
(*b*) Schiff's reagent

Staining method

(*a*) Bring all sections to water

(*b*) Stain in Alcian Blue solution 5 min
(*c*) Wash in distilled water
(*d*) Oxidise in 1 per cent periodic acid 5 min
(*e*) Wash well in distilled water
(*f*) Place in Schiff's reagent 8 min
(*g*) Wash well in running tap water 10 min
(*h*) Counterstain lightly in Mayer's haematoxylin if required
(*i*) Differentiate in 1 per cent acid alcohol, blue, etc.
(*j*) Dehydrate through graded alcohols to xylene
(*k*) Mount in DPX

Results

Acid mucosubstances: blue
Neutral mucosubstances: red
Mixtures: purple

Remarks

(*a*) A good method to separate acid and neutral mucosubstances
(*b*) As Cook states, the method if negative can be taken to mean that a given substance is unlikely to be a mucosubstance.

Method 12.10

Sulphated and carboxylated mucosubstances:
Aldehyde Fuchsin-Alcian Blue

Reagents required

(*a*) Basic fuchsin
(*b*) Conc. HCl
(*c*) Paraldehyde
(*d*) Alcian Blue
(*e*) Acetic acid

Staining solutions

(*a*) Aldehyde fuchsin stain

Basic fuchsin	1 g
60 per cent alcohol	100 ml
Hydrochloric acid (conc.)	1 ml
Paraldehyde	2 ml

Dissolve the dye in the alcohol, then add the acid and paraldehyde. Leave until stain is deep purple, 3-4 days.

(*b*) Alcian Blue

Alcian Blue	1 g
3 per cent acetic acid	100 ml

Staining method

(*a*) Bring all sections to water

(*b*) Stain in aldehyde fuchsin stain — 20 min

(*c*) Rinse well in 70 per cent alcohol

(*d*) Wash in tap water

(*e*) Stain in Alcian Blue solution — 5 min

(*f*) Wash in tap water

(*g*) Dehydrate through graded alcohols to xylene

(*h*) Mount in DPX

Results

Strongly sulphated mucosubstances: deep purple

Weakly sulphated mucosubstances: purple

Carboxyl mucosubstances : blue

Remarks

(*a*) Aldehyde fuchsin is not specific for sulphated mucosubstances

(*b*) The aldehyde fuchsin is a capricious stain that deteriorates and should be replaced every 2 months. Store at 4°C.

Method 12.11

Acid mucosubstances: Dialysed Iron method

Reagents required

(*a*) Ferric chloride

(*b*) Acetic acid

(*c*) Hydrochloric acid

(*d*) Potassium ferrocyanide

(*e*) Distilled water

Preparation of solutions

(*a*) Stock colloidal iron solution

29 per cent ferric chloride	2.2 ml
Distilled water	125 ml

Pour the ferric chloride into boiling distilled water and stir well. The solution will turn dark red; at this point it is removed from the heat and allowed to cool. It is important that the solution is boiling when the ferric chloride is added.

(*b*) Staining solution of collodial iron

	Glacial acetic acid	5 ml
	Distilled water	15 ml
	Stock colloidal iron solution 20 ml	
(*c*)	Acid ferrocyanide mixture	
	Potassium ferrocyanide	2 g
	Hydrochloric acid (conc.)	2 ml
	Distilled water	98 ml

The potassium ferrocyanide is dissolved in the distilled water and the hydrochloric acid is added to this solution.

Sections

All types

Staining methods

(*a*)	Bring all sections to water	
(*b*)	Rinse in 12 per cent acetic acid	
(*c*)	Stain in solution (*b*), collodial iron solution	1 h
(*d*)	Rinse in 12 per cent acetic acid, 4 changes, 3 min each	
(*e*)	Treat section with acid ferrocyanide, solution (*c*)	20 min
(*f*)	Wash in distilled water	
(*g*)	Wash in tap water	5 min
(*h*)	Counterstain in Mayer's Carmalum	10 min
(*i*)	Wash in tap water	
(*j*)	Dehydrate through graded alcohols	
(*k*)	To xylene and mount in DPX	

Results

Acid mucosubstances: bright blue

Nuclei : red

Remarks

Some workers prefer to use the Feulgen reaction as a counterstain.

Method 12.12

Acid mucosubstances: Azure 'A'

Reagents required

(*a*) 0.2 per cent Azure A (aqueous)

(*b*) 5 per cent oxalic acid (aqueous)

(*c*) 1 per cent potassium permanganate (aqueous)

(*d*) 0.2 per cent uranyl nitrate (aqueous)

Sections

All types

Staining method

(*a*)	Bring all sections to water	
(*b*)	Treat with 1 per cent potassium permanganate	5 min
(*c*)	Wash in tap water	
(*d*)	Bleach in 5 per cent oxalic acid	
(*e*)	Wash well in running tap water	
(*f*)	Stain in 0.2 per cent Azure A	5 min
(*g*)	Rinse briefly in water	
(*h*)	Differentiate in 0.2 per cent uranyl nitrate	10-20s
(*i*)	Wash briefly in water, blot dry	
(*j*)	Dehydrate through graded alcohols to xylene	
(*k*)	Mount in DPX	

Results

Acid mucosubstances : purple-red

Nuclei, backgrounds : blue

Remarks

The staining obtained with this method is reasonably alcohol-fast.

Method 12.13

Acid mucosubstances: Toluidine Blue method

Reagents required

(*a*) Toluidine Blue

(*b*) Absolute alcohol

(*c*) Distilled water

Staining solution

Toluidine Blue	100 mg
Absolute alcohol	30 ml
Distilled water	70 ml

Sections

Frozen sections (formalin-fixed)
Cryostat post-fixed
Cryostat pre-fixed
Paraffin sections
Freeze dried

Staining method

(*a*) Bring all sections to water

(*b*) Stain in Toluidine Blue solution 15 min

(*c*) Wash one section in distilled water and mount in glycerin jelly

(*d*) Rinse another section in 95 per cent alcohol

(*e*) Rinse in absolute alcohol

(*f*) Clear in xylene and mount in DPX

Results

Acid mucosubstances : pink (metachromatic)

Nuclei : blue

Remarks

A comparison is made between the two slides, *see* text.

Azure A can be used instead of Toluidine Blue and is preferred by some workers.

Method 12.14

Low Iron Diamine method for acid mucosubstances

Reagents required

N,N-dimethyl-m-phenylenediamine dihydrochloride

N,N-dimethyl-p-phenylenediamine dihydrochloride

40 per cent ferric chloride

3 per cent acetic acid

Alcian Blue

Preparation of staining solutions

(*a*) *Diamine solution*

N,N-dimethyl-m-phenylenediamine dihydrochloride 30 mg

N,N-dimethyl-p-phenylenediamine dihydrochloride 5 mg

Dissolve the salts in 50 ml of distilled water and add 0·5 ml of 40 per cent ferric chloride. It must be freshly prepared or the method will fail.

(*b*) Alcian Blue pH 2.5

Sections

All types

Staining method

(*a*) Bring sections to water

(*b*) Stain in diamine solution 18 h

(*c*) Rinse rapidly in distilled water
(*d*) Stain in Alcian Blue solution 5 min
(*e*) Briefly rinse in tap water
(*f*) Dehydrate through graded alcohols
(*g*) Mount in DPX

Results

Sulphated and non-sulphated mucosubstances: black
Some acidic non-sulphated mucosubstances : blue

Remarks

Neutral mucosubstances may be demonstrated by this method and by the high diamine method, if an additional section is oxidised in 1 per cent periodic acid followed by a wash in tap water before staining in the diamine solution. Neutral mucosubstances in the oxidised section will stain purple-grey.

Method 12.15

High Iron Diamine method for sulphated mucosubstances

Reagents required

(*a*) N,N-dimethyl-m-phenylenediamine dihydrochloride
(*b*) N,N-dimethyl-p-phenylenediamine dihydrochloride
(*c*) 40 per cent ferric chloride
(*d*) Alcian Blue
(*e*) Acetic acid

Preparation of solutions

(*a*) *Diamine solution*

N,N-dimethyl-m-phenylenediamine dihydrochloride 120 mg
N,N-dimethyl-p-phenylenediamine dihydrochloride 20 mg

Dissolve the salts in 50 ml of distilled water then add 40 per cent ferric chloride 1.4 ml. Check pH of solution which should be 1.5. The solution must be freshly prepared.

(*b*) Alcian Blue solution pH 2·5

Sections

All types

Staining method

(*a*) Bring all sections to water

(*b*) Stain in diamine solution for 18 h
(*c*) Wash in water
(*d*) Stain in Alcian Blue solution for 5 min
(*e*) Wash in water
(*f*) Dehydrate through graded alcohols to xylene
(*g*) Mount in DPX

Results

Sulphated mucosubstances: black
Non-sulphated mucosubstances : blue

Method 12.16

Blocking of Aldehyde Groups by Acetylation

Reagents required

(*a*) Acetic anhydride
(*b*) Dry pyridine
(*c*) Potassium hydroxide
(*d*) Periodic acid
(*e*) Schiff's reagent

Preparation of solutions

(*a*) Acetic anhydride solution

Acetic anhydride	16 ml
Dry pyridine	24 ml

(*b*) Potassium hydroxide solution

Potassium hydroxide	1 g
Absolute alcohol	70 ml
Distilled water	30 ml

(*c*) 1 per cent periodic acid
(*d*) Schiff's reagent (see text)

Sections

All types

Suitable control sections

Previous known positive control

Method

(*a*) Bring three sections to water, label 1, 2 and 3
(*b*) Place sections numbered 1 and 2 in acetic anhydride solution 1-24 h. Leave section 3 in distilled water

(*c*) Rinse sections 1 and 2 in distilled water

(*d*) Treat section labelled 2 in KOH solution for 30 min

(*e*) Rinse section 2 in distilled water

(*f*) Apply Schiff's routine to all sections (*see* Method 12.4)

Results

A positive result with sections numbered 1 and 3 with a negative result with no. 2 indicates the reaction was due to 1 :2 glycol groups.

Method 12.17

'Mild'/Methylation for Blocking of Carboxylated Mucins

Reagents required

(*a*) Hydrochloric acid

(*b*) Methanol

(*c*) Alcian Blue

(*d*) Acetic acid

Preparation of staining solutions

(*a*) Methylation solution

Methanol	50 ml
Hydrochloric acid (conc.)	0.4 ml

(*b*) Alcian Blue pH 2.5 (*see text*)

Sections

All types

Suitable control sections

2 previous known positive controls sections and 2 test sections

Method

(*a*) Bring two positive and two control sections to water

(*b*) Treat one positive and one control section with methylation solution at 37°C for 4 h. Leave other two sections in distilled water at 37°C for 4 h

(*c*) Wash all sections in tap water

(*d*) Stain in Alcian Blue solution 5 min

(*e*) Wash in tap water

(*f*) Dehydrate through graded alcohols to xylene and

(*g*) Mount in DPX

Results

Non-sulphated carboxylated mucosubstances: unstained

Sulphated mucosubstances: blue

Remarks

Untreated positive control should be stained.

Method 12.18

Methylation and Saponification-Alcian Blue for carboxy lated mucosubstances

Reagents required

(*a*) Methanol
(*b*) Hydrochloric acid
(*c*) Potassium hydroxide
(*d*) Alcian Blue
(*e*) Acetic acid

Preparation of solutions

(*a*) *Methylating solution* (*see* previous method)

(*b*) *Saponification solution*

Potassium hydroxide	1 g
Absolute alcohol	70 ml
Distilled water	30 ml

(*c*) Alcian Blue solution pH 2.5 (see text)

Sections

All types

Suitable control sections

3 test sections and 3 control sections required

Method

(*a*) Clearly number both test and control slides 1, 2 and 3
(*b*) Bring all sections to 70 per cent alcohol and celloidinise
(*c*) Down to water
(*d*) Place the test and control sections numbered 1 and 2 in methylating solution at 60°C for 5 h. The sections labelled 3 are placed in distilled water at 60°C for 5 h
(*e*) Wash all the sections in tap water 5 min
(*f*) Treat sections labelled 1 with saponification solution for 30 min. Leave sections numbered 2 and 3 in 70 per cent alcohol for 30 min
(*g*) Wash sections in water 5 min
(*h*) Place in 70 per cent alcohol

(*i*) Remove celloidin in alcohol and ether 50:50
(*j*) Rinse in 70 per cent alcohol
(*k*) Wash in water
(*l*) Stain all sections in Alcian Blue solution 5 min
(*m*) Rinse in tap water
(*n*) Use suitable background counterstain
(*o*) Wash in water
(*p*) Dehydrate through graded alcohols to xylene
(*q*) Mount in DPX

Results

Sections numbered 1 = carboxylated mucosubstances : blue

Sections numbered 2 = sulphated and carboxylated mucosubstances : blocked, no blue staining

Sections numbered 3 = all acid mucosubstances : blue

Remarks

(*a*) On some occasions this method works well, on others it appears capricious
(*b*) During saponification sections will tend to float off the sides.

Method 12.19

Hyaluronidase digestion: Alcian Blue

Reagents required

(*a*) Testicular hyaluronidase
(*b*) Phosphate buffer pH 6.7
(*c*) Alcian Blue solution

Preparation of solutions

(*a*) Hyaluronidase solution
Testicular hyaluronidase (600 units/mg) 10 mg
Phosphate buffer pH 6.7 10 ml (*see text*)
(*b*) Alcian Blue solution either pH 2.5 or pH 0.5 (see Remarks)

Sections

All types

Suitable control sections

Known positive containing hyaluronic acid or chondroitin sulphates

Staining method

(*a*) Bring two test and two control sections to distilled water (*b*) Treat one test and one control section with hyaluronidase solution at 37°C for 3 h. The other two sections are placed in buffer solution only, at 37°C for 3 h

(*c*) Wash all sections in water

(*d*) Stain in Alcian Blue solution 5 min

(*e*) Wash in water

(*f*) Counterstain in Mayer's Carmalum 5 min

(*g*) Wash in water

(*h*) Dehydrate through graded alcohols to xylene

(*i*) Mount in DPX

Results

Hyaluronic acid and chondroitin sulphates: negative

Other acid mucosubstances: blue

Nuclei : red

Remarks

(*a*) Loss of staining in the test section after digestion compared with the section incubated in buffer only indicates the presence of one or more of the hyaluronidase labile mucosubstances.

(*b*) If Alcian Blue at pH 0.5 is employed, any highly sulphated mucosubstances will be shown.

Method 12.20

Sialidase digestion: Alcian Blue

Reagents required

(*a*) Sialidase

(*b*) Acetate buffer pH 5.5

(*c*) Calcium chloride

(*d*) Alcian Blue solution

Preparation of solutions

(*a*) ***Sialidase***

Neuraminidase available ready prepared from Koch-Light Ltd.	1 ml
Acetate buffer, pH 5·5	4 ml
Calcium chloride	50 mg

(*b*) Alcian Blue solution pH 2.5 (See text)

Sections

All types

Suitable control sections

Known positive containing sialic acid

Staining method

(*a*) Bring 2 test and 2 control sections to water

(*b*) Treat 1 test and 1 control section with sialidase solution at 37°C for 18 h. Incubate the other 2 sections with buffer solution only at 37°C for 18 h

(*c*) Wash well in tap water

(*d*) Stain in Alcian Blue pH 2-5 for 5 min

(*e*) Wash in water

(*f*) Counterstain in Mayer's Carmalum

(*g*) Wash in water

(*h*) Dehydrate through graded alcohols to xylene and

(*i*) Mount in DPX

Results

Sialidase labile sialomucins: unstained
Other acid mucosubstances: blue
Nuclei : red

13

PROTEINS AND AMINO ACIDS

Proteins are the third of the components that go to make up cell *cytoplasm*, the other two being *lipids* and *carbohydrates*. Similar problems of identification apply to the proteins as to the other two, mainly because of the *intermingling* of all three components. Proteins are to be found in all cells and tissues. The actual demonstration of protein material is not difficult, but the identification of any particular protein is.

Proteins may be very simply classified as shown in Table elsewhere in this chapter.

Table 13.1

Simple proteins	*Conjugated proteins*
Albumins	Lipoproteins
Globulins	Mucoproteins
Fibrous proteins	Glycoproteins
Histones and others	Nucleoproteins

The simple proteins are the naturally occurring proteins that upon hydrolysis will yield amino acids only. These can be divided into fibrous proteins (e.g. *collagen*, *keratin* and *fibrin*), and globular proteins (e.g. *albumin*, *globulin*). The first group are mainly insoluble in water, but the second are freely soluble. Because of their insolubility, it is much easier to demonstrate the *fibrous proteins* than the globular ones.

The conjugated proteins consist of simple proteins combined with a non-protein material, for example, lipids in lipoproteins and nucleic acids in nucleoproteins.

DEMONSTRATION OF FIBROUS PROTEINS

The commonly occurring fibrous proteins such as collagen, reticulin, elastin, *fibrin/fibrinoid* and keratin are easily demonstrated using routine *histopathological* staining methods. All, with the exceptions of *reticulin*, are easily seen in haematoxylin and eosin stained sections, and are usually strongly eosinophilic. *Collagen* can be demonstrated by the red acid fuchsin in the Van Gieson stain, and by certain components of the various trichrome stains.

The finer fibres of reticulin are demonstrated by silver reduction methods, and elastic fibres can be stained by resorcin or orcein. Fibrin/fibrinoid, derived from circulating plasma proteins, is stained dark-blue by Mallory's PTAH method and bright red by the MSB technique developed by Lendrum *et al.* Because fibrin/fibrinoid has a high tryptophan content it can also be demonstrated in tissues by the DMAB-nitrite method of Adams.

Keratin has a particularly high content of the sulphur containing *amino acid*, *cystine*, and the *keratin* which forms hair is particularly richly endowed with this amino acid. Mature keratin will therefore be demonstrable by the DDD or *Mercury Orange methods* which depend on the presence of *disulphide linkages*.

Methods for Proteins and Amino Acids

The methods discussed below do not demonstrate the protein molecule as a whole, but are dependent on the presence of certain types of reactive groups or special types of linkages within the protein molecule. These groups or linkages reside within certain of the amino acids making up the protein, and so these methods enable some conclusion to be drawn about the amino acids present in the tissue section.

Unfortunately this whole segment of amino-acid histochemistry is confused because of the uncertainty about the specificity of the methods used. Histochemical reactions are available for some, but not all, of the reactive groups of the protein molecule.

The groups and linkages for which histochemical methods exist are enumerated below:

(*a*) Protein-bound amino groups (e.g. lysine)

Methods 13.1, 13.2, 13.3

(*b*) Disulphide and sulphydryl linkages

(e.g. in cystine and cysteine) Methods 13.4, 13.5, 13.6, 13.7

(*c*) Guanidyl groups (e.g. in arginine) Method 13.8

(*d*) Indole groups. (e.g. tryptophan and tryptamine) Method 13.9

(*e*) Phenyl groups (e.g. tyrosine) Methods 13.10, 13.11

Protein-bound Amino Groups

Amino groups of lysine, ornithine and terminal peptides cannot be demonstrated individually, but the group as a whole may be demonstrated. The simplest and most reliable methods are the *Ninhydrin-Schiff* method of Yasuma and Ichikawa, the DNFB (*dinitrofluorobenzene*) method of Sanger, modified by both Danielli and Burstone, and the hydroxynaphthaldehyde method of Weiss, Tsou and Seligman.

The DNFB method performed in the manner given in Method 13.2 demonstrates *sulphydryl* groups and the hydroxyphenyl groups of *tyrosine*, as well as the protein-bound amino groups. Nevertheless this does not detract from its value as a general protein stain; various blocking techniques enable the method to be used more specifically if desired.

Disulphide and Sulphydryl Linkages

These sulphur-containing groups are found in the amino acids cystine, cysteine and *methionine*. Reduction of the *disulphide* linkage (-S-S) will give the sulphydryl group (-S-H). Lillie, stated that great caution must be exercised when distinguishing between the two groups.

The best methods available for the demonstration of the sulphydryl and disulphide groups are the *dihydroxydinaphthyldisulphide* method of Barrnett and Seligman (the DDD method) and the Mercury Orange method of Bennett and Watts. The demonstration of disulphide groups with the former method depends upon the reduction of these groups to sulphydryl. (*See* Method 13.5.)

Alternatively, the performic acid-alcian blue method of Adams and Sloper may be used for disulphide groups, but it is less sensitive than the others.

Guanidyl Groups

The only amino acid containing the guanidyl group which is demonstrable histochemically in human tissue is arginine. The histochemical method is based on the well known Sakaguchi reaction, in which a red colour develops when arginine reacts with a-naphthol and hypochlorite in an alkaline solution. The original biochemical reaction was modified by Baker for histochemical use; the orange-red colour which develops is transient, so the preparation must be examined immediately. To obtain satisfactory results the tissue must contain a high concentration of arginine; testis is particularly rich in arginine.

Indole Groups

The indole-containing amino acids which can be demonstrated histochemically are tryptophan and tryptamine. Of the many methods available the most reliable is the DMAB-nitrite method of Adams. The best results are obtained using freeze dried sections which have been fixed in formalin vapour; sites of high concentration of typtophan show an intense blue colour.

Phenyl Groups

The only amino acid which contains the hydroxyphenyl group and which can be demonstrated histochemically is tyrosine. Because tyrosine is such a very frequent component of almost all tissue proteins, the methods for tyrosine can be regarded as demonstrating protein in general. The histochemical methods available are the Millon reaction modified by Baker, and the diazonium coupling method of Glenner and Lillie.

STAINING METHODS

Ninhydrin-Schiff for Amino Groups

Ninhydrin at neutral pH will react with x-amino groups. Yasuma and Ichikawa published the first histochemical technique that involved the use of ninhydrin. The α-amino group reacts with ninhydrin and yields an aldehyde. This is then demonstrated by Schiff's reagent.

The precise nature of the reaction is not known although the results appear to be specific. Ninhydrin is normally used as an 0.5 per cent solution in absolute alcohol and at 37°C. Fixation is best carried out in neutral formol saline.

DNFB method for Amino,
Sulphydryl and Hydroxy-phenyl Groups

Dinitrofluorobenzene (DNFB) has frequently been used as a reagent for protein end groups. It reacts with free amino groups in proteins, the amino groups of lysine and *hydroxylysine*, the *hydroxyphenyl* groups of tyrosine and the sulphydryl groups of cysteine. The end product of the DNFB-amino group reaction is coloured yellow, whereas the product of the DNFB-tyrosine reaction is colourless.

This colourless product can be made visible by *diazotisation* and subsequent coupling with a phenol or aromatic amine. The most suitable coupling agent is '*H-acid*' (8-amino-l-naphthol-3: 6-disulphonic acid) which gives an insoluble reddish-purple deposit as a final reaction product. Burstone considered that the DNFB-Hacid combination could be satisfactorily used to demonstrate primary amino groups (e.g. lysine) and hydroxy groups (e.g. *tyrosine*). More specific separation of staining due to amino, sulphydryl and aromatic hydroxyl groups can be obtained by using specific blocking agents before performing the DNFB reaction. This problem is dealt with by Pearse.

Hydroxynaphthaldehyde Methods for Amino Groups

Weiss, Tsou and Seligman suggested the use of 3-hydroxy2-naphthaldehyde for the demonstration of protein-bound amino groups at pH 8.5. The primary reaction product is post-coupled with a *diazonium* salt at pH 7.4 to form a blue colour.

Formalin fixation may partially suppress the reaction, although a positive result may still be obtained after routine formalin fixation provided that there are a large number of reactive amino groups.

The dihydroxyldinaphthyldisulphide Method for S-S and S-H Groups

This method, known as the DDD reaction, was published by Barrnett and Seligman. The method with modifications can be used to demonstrate: (*i*) sulphydryl groups and disulphide linkages; (*ii*) sulphydryl groups only; and (*iii*) disulphide linkages. The methods are basically the same. For the first method, thioglycollic acid splits the disulphide linkages and forms sulphydryl groups.

These, together with any pre-existing sulphydryl groups, are then demonstrated. Both types of sulphydryl groups react with the DDD reagent and the primary reaction product, which is colourless, is then coupled with the diazonium salt fast blue B. The colour of the final reaction product is blue.

The method is considered by many authors to be very specific, but unfortunately, the intensity of the positive blue colour is not very strong. If the reaction is required for the demonstration of preexisting sulphydryl groups only, the thioglycollic acid stage is omitted, and the disulphide linkages are not then demonstrated since the bond is not broken.

To demonstrate disulphide linkages only, it is necessary to use N-ethyl maleimide to block the existing sulphydryl groups. The disulphide linkages are then reduced to sulphydryl groups and demonstrated by the method. Barrnett and Seligman stated that a weak reducing agent must be used, otherwise the N-ethyl maleimide blocking of the sulphydryl may be reversed. They recommended 10 per cent potassium cyanide for this purpose.

The Mercury Orange Method for S-H Groups

This was published by Bennett and Watts. The reagent used is now called Mercury Orange. It was originally called Red Sulphydryl Reagent (R.S.R.) but this name has now passed out of use. The method is said to be specific for sulphydryl groups. The sulphydryl groups react with Mercury Orange to form a mercaptide, which is coloured and visible with the light microscope. The colour however, is not intense.

Performic Acid-Alcian Blue Method for S-S Groups

This technique was introduced by *Adams and Sloper* for the demonstration of cysteine in paraffin sections. The technique can also be applied to cryostat and freeze dried material, the latter giving the best results of the three. The authors used Pearse's performic acid (freshly made daily), to oxidise the cysteine to cysteine sulphuric acid. This was subsequently stained using the basic dye, Alcian Blue.

It is important that the pH of the Alcian Blue solution should not be less than 0.2 otherwise the localisation is unsatisfactory. Adams stated that for disulphide groups (-S-S) and sulphydryl groups (-SH), the specificity and selectivity of the method are good.

Sakaguchi Method for Arginine

It is difficult to obtain consistent results using this technique. The intensity of the final colour is sometimes weak and is unstable, so sections must be examined immediately. Various modifications have appeared since the original paper. In his paper, Sakaguchi described the reaction of α-naphthol with arginine in the presence of strong alkali, to give a red colour.

Baker's modification of this method is probably the best. Other naphthols have been tried and in some cases a more stable colour reaction can be obtained; among these may be mentioned 8-hydroxyquinoline and 2:4-dichloro-a-naphthol. It is important that the sections used for this method are not too thin; satisfactory results are obtained using sections of about 15 microns. They show the weak colour reaction better and are more able to withstand the technique which is very destructive to sections.

The DMAB-nitrite Method for Tryptophan

Histochemical techniques for the demonstration of tryptophan have been given by Glenner, Adams and Glenner and Lillie. The methods described by these authors employ p-dimethylaminobenzaldehyde (DMAB).

The difference between the Glenner and Adams techniques is in the composition of the oxidizing solutions. The Glenner and Lillie method is a post-coupling technique.

The method of Adams is recommended because it is easy to carry out and the results are usually reproducible. In the Adams method, the tryptophan reacts with p-*dimethylaminob-enzaldehyde* to produce a compound known as 8-carboline. Oxidation of this product by nitrite solution produces a blue pigment.

The Millon Reaction for Tyrosine

This reaction, known to biochemists for many years, has been modified for histochemical purposes. The two best and most popular methods are those of Bensley and Gershand of Baker. The method, which is specific for tyrosine, has one disadvantage in that it gives a weak colour reaction.

The recommended method, and the one considered here, is a modification of Baker's. It was shown by Meyer that if *mercuric chloride* is applied to *tyrosine* in the presence of potassium nitrite in an acid solution, a red colour is produced. The various modifications of the method have centred around the choice of the mercuric salt to be used. Baker found that *Folin's reagent* gave him the strongest reaction. In this reagent, *mercuric sulphate* is used with sulphuric acid.

Baker also experimented with fixation, and found that formaldehyde-fixed tissue gave stronger colouration than tissues treated with other routine fixatives. To produce a satisfactory colour reaction it is necessary to heat the mercuric sulphate solution.

Generally speaking, the higher the temperature, the more intense the colour. Baker heated celloidin sections to above 90°C. Other types of sections are heated to 70°C, when a reddish pink colour may be obtained at the sites of tyrosine. Alternative methods for the demonstration of tyrosine are the diazotisation coupling method of Glenner and Lillie and the DNFB method of Danielli.

Diazotisation-coupling Method for Tyrosine

This method, developed by Glenner and Lillie in 1959, was based on earlier work by Lillie who claimed that prolonged nitrosation of tyrosine led to the formation of diazonium nitrates. These diazonium nitrates could subsequently be coupled with amines in alkaline solution to give coloured products. The histochemical method uses 'S-acid' (8 amino 1-naphthol-5-sulphonic acid) as the coupling amine and the incubations are carried out in the dark.

FIXATIVES

For most purposes *formalin fixation* is satisfactory for protein histochemistry, despite the fact that formalin modifies many of the reactive groups in the *protein molecules*. This does not interfere with most of the protein methods, since adequate washing in water after the formalin fixation will leave enough of the active groups sufficiently unchanged to react normally with the histochemical reagent.

However, a non-formalin fixative is preferred for the *hydroxynaphthaldehyde* method for amino groups. Barnett and Roth tested a series of fixatives in certain protein *histochemical* methods and found that only osmium tetroxide completely inhibited any of the reactions.

In general, freeze dried sections give the best staining intensity and localisation in protein histochemistry, although the warm *formalin vapour* used for fixation tends to interfere with the reactive groups more than routine formalin solutions.

Method 13.1

Amino groups: The Ninhydrin-Schiff method

Reagents required

(*a*) Ninhydrin
(*b*) Absolute alcohol
(*c*) Schiff's reagent

Preparation of solutions

(*a*) Ninhydrin 0.5 per cent solution

Ninhydrin	500 mg
Absolute alcohol	100 ml

(*b*) Schiff's reagent

Sections

Freeze dried
Paraffin sections
Cryostat unfixed
Cryostat prefixed

Suitable control tissue

Pancreas

Method

(*a*) Bring sections down to 70 per cent alcohol
(*b*) Treat with solution (*a*), ninhydrin, at 37°C overnight

(*c*) Wash in running tap water
(*d*) Place sections in Schiff's reagent, solution (*b*), for 45 min
(*e*) Wash well in running tap water
(*f*) Counterstain if required in haematoxylin
(*g*) Wash in tap water
(*h*) Dehydrate through graded alcohols to xylene and mount

Results

α-Amino groups : pink to red

Remarks

(*a*) 1 per cent Alloxan, also in absolute alcohol, may be used instead of 0.5 per cent ninhydrin.
(*b*) Control sections may be needed for other PAS-positive material.

Method 13.2

DNFB method for tyrosine, SH and NH, groups

Reagents required

(*a*) 2-4 Dinitrofluorobenzene
(*b*) Ethyl alcohol
(*c*) Sodium hydrogen carbonate
(*d*) Sodium hydrosulphite
(*e*) Sodium nitrite
(*f*) Hydrochloric acid
(*g*) H-acid (8 amino-l-naphthol 3,6-disulphonic acid)
(*h*) Veronal acetate buffer pH 9.4

Preparation of solutions

(*a*) Dinitrofluorobensene solution

2-4 dinitrofluorobenzene saturated in 90 per cent ethyl alcohol saturated with sodium hydrogen carbonate.

(*b*) Nitrous acid

5 per cent sodium nitrite	10 ml
2N hydrochloric acid	40 ml

(*c*) 'H'-acid

H-acid saturated in 0.1M veronal acetate buffer pH 9.4 (*see text*).

Sections

All types

Suitable control sections

Pancreas

Method

(*a*) Bring sections to absolute alcohol and allow to dry in air
(*b*) Place in DNFB solution 2-16 h (overnight is usually ideal)
(*c*) Rinse in 90 per cent alcohol (3 changes) then tap water
(*d*) Treat with 5 per cent sodium hydrosulphite at 45°C for 30 min
(*e*) Wash in distilled water
(*f*) Treat sections with nitrous acid solution at 4°C for 30 min
(*g*) Wash in cold distilled water
(*h*) Treat with H-acid solution at 4°C for 15 min
(*i*) Wash in tap water
(*j*) Dehydrate through graded alcohols to xylene
(*k*) Mount in DPX

Results

A positive reaction is reddish purple

Method 13.3

Hydroxynaphthaldehyde method: NH, groups

Reagents required

(*a*) 3-Hydroxy-2-naphthaldehyde
(*b*) Acetone
(*c*) Veronal acetate buffer pH 7.4, 8.5
(*d*) Fast Blue B

Preparation of solutions

(*a*) Incubating solution

3-Hydroxy-2-naphthaldehyde 20 mg Acetone	20 ml
0.1 M veronal acetate buffer (pH 8.5)	30 ml

The 3-hydroxy-2-naphthaldehyde is dissolved in the acetone and the buffer added.

(*b*) Diazonium salt solution

This is prepared during the technique using Fast Blue B and veronal acetate buffer pH 7.4.

Sections

All types (non-formalin fixation preferred).

Suitable control sections

Pancreas

Method

(*a*) Bring sections to water
(*b*) Place sections in incubating solution for 1 h at room temperature
(*c*) Wash well in 3 changes of distilled water
(*d*) Place sections in 0.1M veronal acetate buffer pH 7.4, add to this solution 25 mg of Fast Blue B and shake solution. Leave sections in this solution for 5 min
(*e*) Wash well in tap water
(*f*) Dehydrate through graded alcohols to xylene
(*g*) Mount in DPX

Result

Reactive NH, groups: blue

Method 13.4

Disulphides: DDD reaction

Reagents required

(*a*) 2 :2'-Dihydroxy-6 :6'-dinaphthyl disulphide (DDD)
(*b*) Absolute alcohol
(*c*) Acetic acid
(*d*) Veronal acetate buffer, pH 8·5 (0·1 M)
(*e*) Absolute ether
(*f*) Fast Blue B
(*g*) Phosphate buffer, pH 7·4 (0·1 M)
(*h*) 0·5 per cent Celloidin
(*i*) N-Ethyl-maleimide
(*j*) Sodium hydroxide (0·1 N)
(*k*) Potassium cyanide

Preparation of solutions

(*a*) Blocking solution

N-Ethyl-maleimide	1.25 g
0.1 M phosphate buffer, pH 7.4	100 ml

(*b*) Reducing agent

Potassium cyanide	10 ml
Distilled water	90 ml

(*c*) Incubating solution

DDD reagent	25 mg dissolved in
Absolute alcohol	15 ml
0.1 M Veronal acetate buffer	35 ml

(*d*) Fast Blue B solution

Fast Blue B	50 mg
0.1 M Phosphate buffer, pH 7.4	50 ml

This solution must be freshly prepared before use.

Sections

Freeze dried

Paraffin sections

Cryostat sections

Suitable control sections

Pancreas, pituitary, skin

Method

(*a*) Dewax paraffin sections

(*b*) Bring all sections to 85 per cent alcohol

(*c*) Coat sections in 0·5 per cent celloidin

(*d*) Rinse in 70 per cent alcohol

(*e*) Wash in tap water

(*f*) Place sections in blocking agent, solution (*a*), at 37°C for 4 h

(*g*) Wash sections in 1 per cent acetic acid for 2 min

(*h*) Rinse in tap water

(*i*) Place sections in reducing agent (potassium cyanide), solution (*b*), for 2 h at 60°C

(*j*) Wash in tap water

(*k*) Incubate sections in solution (*c*), DDD reagent, at 50°C for lh

(*l*) *Allow* to cool to room temperature

(*m*) Wash sections in distilled water

(*n*) Place sections in distilled water, acidified with acetic acid, pH 4·5, 5 min

(*o*) Place sections in fresh distilled water with acetic acid, pH 4·5, 5 min

(*p*) Rinse sections in 70 per cent alcohol 2 min

(*q*) Rinse sections in 90 per cent alcohol 2 min

(r) Rinse sections in absolute alcohol 2 min
(s) Rinse sections in absolute ether 2 min
(t) Rinse sections in absolute alcohol 2 min
(u) Rinse sections in 90 per cent alcohol 2 min
(v) Rinse sections in 70 per cent alcohol 2 min
(w) Rinse sections in distilled water 2 min
(x) Stain sections in solution (d) Fast Blue B 2 min
(y) Wash sections in running tap water.
(z) Dehydrate through graded alcohols to xylene and mount.

Results

Disulphides : bluish-reddish violet

Remarks

This is an indirect result after the sulphydryl groups have been blocked by the first solution. The disulphides are reduced to sulphydryl and then demonstrated. Potassium cyanide is used as the reducing agent to avoid unblocking the sulphydryl groups.

The Fast Blue B solution must be prepared freshly just before use.

Method 13.5

Sulphydryl groups : DDD reaction

Reagents required

(a) 2 :2'-Dihydroxy-6 :6'-dinaphthyl disulphide (DDD)
(b) Absolute alcohol
(c) Acetic acid
(d) Veronal acetate buffer, pH 8.5
(e) Ether absolute
(f) Fast Blue B
(g) Phosphate buffer, pH 7.4

Preparation of solutions

(a) Incubating solution (DDD)

0.1 M Veronal acetate buffer, pH 8.5	35 ml
DDD reagent	25 mg dissolved in
Absolute alcohol	15 ml

(b) Fast Blue B solution

Fast Blue B	50 mg
Phosphate buffer pH 7.4 (0.1 M)	50 ml

This solution is freshly prepared just before use.

Sections

Freeze dried
Paraffin sections
Cryostat sections

Suitable control sections

Pancreas, pituitary, skin, trachea

Method

(*a*)	Bring all sections to water	
(*b*)	Incubate sections in solution (*a*), DDD reagent, at 50°C for lh	
(*c*)	Allow to cool to room temperature	
(*d*)	Wash sections in distilled water	
(*e*)	Place sections in distilled water acidified with acetic acid, pH 4.5 for 5 min	
(*f*)	Repeat with fresh acidified distilled water	5 min
(*g*)	Rinse sections in 70 per cent alcohol	2 min
(*h*)	Rinse sections in 95 per cent alcohol	2 min
(*i*)	Rinse sections in absolute ether	2 min
(*j*)	Rinse sections in 95 per cent alcohol	2 min
(*k*)	Rinse sections in 70 per cent alcohol	2 min
(*l*)	Rinse sections in distilled water	2 min
(*m*)	Stain section in solution (*b*), Fast Blue B	2 min
(*n*)	Wash sections in running tap water	
(*o*)	Dehydrate through graded alcohols to xylene and mount in DPX	

Results

Sulphydryl groups : reddish purple

Remarks

The Fast Blue B solution should be freshly prepared before use each time the method is applied.

Method 13.6

Sulphydryl groups: Mercury Orange method

Reagents required

(*a*) Mercury Orange
(*b*) N' :N'-Dimethyl formamide

Preparation of solution

Mercury Orange, saturated solution in N' :N'-dimethyl formamide, 40 ml

Sections

Freeze dried

Paraffin sections

Cryostat

Suitable controls

Pancreas, skin, pituitary

Method

(*a*) Paraffin embed material to absolute alcohol then with

(*b*) Frozen sections, place directly into Mercury Orange solution, 2 days

(*c*) Rinse rapidly in 2 changes of absolute alcohol

(*d*) Place sections in xylene then mount in DPX

Results

Sulphydryl groups : pale orange-orange-red

Remarks

Mercury Orange can also be dissolved in absolute alcohol.

Method 13.7

Disulphides: Performic Acid-Alcian Blue method

Reagents required

(*a*) Formic acid

(*b*) Hydrogen peroxide

(*c*) Sulphuric acid (conc.)

(*d*) Alcian Blue

(*e*) Absolute alcohol

Preparation of solutions

(*a*) Oxidizing solution: performic acid

98 per cent Formic acid	40 ml
100 vol. Hydrogen peroxide	4 ml
Sulphuric acid	0.5 ml

(*b*) Staining solution

Alcian Blue	1 g
98 per cent Sulphuric acid	2.7 ml
Distilled water	47.2 ml

Sections

Freeze dried

Paraffin

Cryostat

Suitable controls

Pituitary, skin

Method

(*a*) Bring all sections to water
(*b*) Remove excess water by blotting
(*c*) Place sections in oxidizing solution 5 min
(*d*) Wash in tap water 10 min
(*e*) Dry section by gently heating to 60°C till just dry
(*f*) Rinse in tap water
(*g*) Stain in Alcian Blue solution at room temperature for 1 h
(*h*) Wash in running tap water
(*i*) Counterstain if required
(*j*) Wash in tap water
(*k*) Dehydrate through graded alcohols to xylene and mount in DPX

Results

Disulphides : dark blue

Smaller amounts of disulphides: light blue

Remarks

The oxidizing solution must be prepared fresh daily. After preparation it should be allowed to stand for I h before use. The section may tend to lift after the oxidizing solution has been applied.

Method 13.8

Arginine: Sakaguchi method

Reagents required

(*a*) Sodium hypochlorite
(*b*) α-Naphthol
(*c*) Sodium hydroxide
(*d*) Pyridine
(*e*) Chloroform
(*f*) 70 per cent Alcohol

Preparation of solutions

Solution (*a*)

1 per cent Sodium hydroxide

Solution (b)

α-Naphthol	1 g
70 per cent Alcohol	100 ml

Solution (c)

Milton	1 ml
Distilled water	99 ml

Milton is the proprietory name of stable sodium hypochlorite.

Incubating solution (d)

Solution (*a*)	2 ml
Solution (*b*)	2 drops
Solution (*c*)	4 drops

Pyridine-chloroform solution (d)

Pyridine	30 ml
Chloroform	10 ml

Sections

Freeze dried

Cryostat

Paraffin sections

Suitable control sections

Testis

Method

(*a*) Bring all sections to water

(*b*) Rinse in 70 per cent alcohol

(*c*) Cover section with incubating solution (*d*) 15 min

(*d*) Drain and blot dry

(*e*) Immerse in pyridine-chloroform, solution (*e*) 2 min

(*f*) Mount in pyridine-chloroform mixture and ring coverslip

Result

Arginine : orange-red

Remarks

The slide should be looked at microscopically immediately.

Method 13.9

Tryptophan: DMAB-nitrite method

Reagents required

(*a*) p-Dimethylaminobenzaldehyde

(*b*) Hydrochloric acid
(*c*) Sodium nitrite
(*d*) Acid alcohol (1 per cent)

Preparation of solutions

(*a*) *p-Dimethylaminobenzaldehyde*

p-Dimethylaminobenzaldehyde	5 g
Hydrochloric acid (conc.)	100 ml

(*b*) *Sodium nitrite*

Sodium nitrite	1 g
Hydrochloric acid (conc.)	100 ml

(*c*) *Acid alcohol*

Hydrochloric acid (conc.)	1 ml
70 per cent Alcohol	99 ml

Sections

Freeze dried
Paraffin sections
Cryostat

Suitable control sections

Pancreas, duodenum, pituitary

Method

(*a*) Bring sections to alcohol
(*b*) Celloidinise in 0·5 per cent celloidin

(*c*) Place sections in solution (*a*), DMAB	1 min
(*d*) Transfer sections to solution (*b*), sodium nitrite	1 min
(*e*) Wash carefully in tap water	30 sec
(*f*) Rinse sections in solution (*c*), acid alcohol	15 sec

(*g*) Dehydrate through graded alcohols to xylene and mount.

Results

Tryptophan: deep blue

Remarks

This method gives good localisation.

Method 13.10

Tyrosine: Millon reaction

Reagents required

(*a*) Mercuric sulphate
(*b*) Sulphuric acid

(*c*) Sodium nitrite

Preparation of solutions

Solution (*a*)

Distilled water	90 ml
Sulphuric acid (cons.)	10 ml

To this solution add 10 g mercuric sulphate and heat until dissolved. Cool to room temperature and add 100 ml distilled water.

Solution (*b*)

Sodium nitrite	250 mg
Distilled water	10 ml

Staining solution (*c*)

Solution 1	10 ml
Solution 2	1 ml

Sections

Freeze dried

Paraffin sections

Cryostat sections

Celloidin sections

Suitable controls

Pancreas, duodenum

Method

(*a*)	Bring all sections to water	
(*b*)	Place sections in a small beaker, add solution (*c*) and boil gently	2 min
(*c*)	Allow to cool to room temperature	
(*d*)	Wash sections in distilled water	2 min
(*e*)	Repeat wash in distilled water	2 min
(*f*)	Repeat wash in distilled water	2 min
(*g*)	Dehydrate through graded alcohols to xylene and mount in DPX	

Result

Tyrosine : red, pink or yellowish red.

Method 13.11

Diazotisation-coupling method for tyrosine

Reagents required

(*a*) Sodium nitrite

(*b*) Acetic acid
(*c*) 8-Amino-l-naphthol-5-sulphonic acid (S-acid)
(*d*) Potassium hydroxide
(*e*) Ammonium sulphamate
(*f*) Hydrochloric acid

Preparation of solutions

(*a*)	Incubating solution	
	Sodium nitrite	6.9 g
	Acetic acid	5.8 ml
	Distilled water	94 ml
(*b*)	8-amino-l-naphthol-5-sulphonic acid	1 g
	Potassium hydroxide	1 g
	Ammonium sulphamate	1 g
	70 per cent alcohol	100 ml

Sections

All types

Suitable control sections

Pancreas

Method

(*a*) Place sections in incubating medium at 4°C overnight in the dark
(*b*) Rinse in distilled water at 4°C
(*c*) Treat with solution (*b*) at 4°C for 1 h also in the dark
(*d*) Rinse in 3 changes of 0.1 NHC 15 min each
(*e*) Wash in tap water 10 min
(*f*) Dehydrate through graded alcohols to xylene
(*g*) Mount in DPX

Results

Tyrosine-containing proteins: purple-red.

14

LIPIDS

The terminology applied to lipids has been ill-defined and complex; in some instances different names have been used to describe the same substance. The term '*lipid*' will be used in this chapter to include all naturally occurring fats and fat-like materials which have formerly been referred to as *lipoid*, *lipin*, *lipine* and *lipide*. The majority of lipids are soluble or part-soluble in organic solvents and for this reason, many of the histochemical techniques are applied to frozen sections.

The means of demonstration of lipid substances includes the use of *lipid dyes*, *microscopy* and extraction techniques. The information obtained from histochemical methods cannot be compared to a biochemical classification due to a lack of specificity of the histochemical techniques. This is due in part to the fact that lipids are frequently found in tissue sections as mixtures of several substances.

They are often bound to proteins (*lipoproteins*) or carbohydrates (*glycolipids*) with the resulting alteration of their chemical and physical properties leading to the failure of techniques such as differential solubility, and the possible misleading result of staining methods for specific lipid substances. Lipids are normal components of all tissues and are usually in the form of stored lipids or as particular lipid structures such as myelin. In certain diseases abnormally large amounts of lipids can be found in cells of different organs of the body and these are known as the lipid storage diseases.

There is no agreement on how lipids should be classified, it is probably easiest to make the following subdivisions.

(*a*) Simple lipids (*b*) Compound lipids

(*c*) Derived lipids

Lipids in general terms can be termed as compounds of long chain fatty acids with an alcohol.

SIMPLE LIPIDS

These are esters of both saturated and unsaturated long chain fatty acids with alcohols. Long chain fatty acids are neutral and therefore this type of lipid is frequently called neutral lipid. The fatty acids can be stearic, palmitic and oleic acids, they will dissolve in organic solvents and are insoluble in water. The neutral lipids can suitably be termed triglycerides.

Also included under the heading of simple lipids are ester waxes. They are the esters of different fatty acids combined with the steroid alcohol, cholesterol. Simple lipids are demonstrated by the Sudan dyes depending upon the preferential absorption of Sudan dyes by the simple lipids (Methods 14.1, 14.2). The red oxazone component of the Nile Blue sulphate technique (Method 14.3) will stain them red.

COMPOUND LIPIDS

These contain a non-lipid group, as well as long chain fatty acids and an alcohol. They may be subdivided into the following groups.

(*a*) Phospholipids

(*b*) Glycolipids

Phospholipids

These on hydrolysis yield a fatty acid, phosphoric acid linked to a nitrogenous base, and an alcohol which is usually glycerol. Phospholipids occur in many tissues, brain, liver and heart muscle being particularly rich. They are regarded as esters of phosphatidic acid and are also known as glycerophosphatides. They can be subdivided into :

(*i*) Lecithin

(*ii*) Kephalins

(*iii*) Plasmals

(*iv*) Sphingomyelin

Lecithin

These phospholipids are the major components of the phospholipids demonstrated in tissue sections. The alcohol is glycerol, to which fatty acids and the phosphoryl base are esterified, and the base is choline.

Kephalins

These are of similar chemical composition with the exception that the organic base is either ethanolamine or serine.

Plasmals

These are found in tissues as two types, firstly as acetal phosphatides and secondly as plasmalogens. They are closely related chemically and are often considered together. They consist of aldehydes of fatty acids usually of palmitic or stearic, incorporated into the phosphatidyl ester structure.

Sphingomyelin

Contain a fatty acid, phosphoric acid, choline and the complex alcohol sphingosine. Apart from their lack of a sugar component sphingomyelins are similar to cerebrosides and in some instances are not classed under the broad heading of phospholipids. They can be histochemically differentiated from the cerebrosides by their failure to react with the PAS reaction, and can be demonstrated by the NaOH OTAN method.

GLYCOLIPIDS

Two major groups of glycolipids occur in tissue sections, firstly, cerebrosides, which contain a single fatty acid chain, a complex alcohol (usually sphingosine) and one or more hexose sugars, usually galactose.

The second group comprises the gangliosides, which are chemically similar to cerebrosides but also contain neuraminic acid or its derivatives. Histochemically the two glycolipids cannot be separated. They can be shown by the Periodic Acid—Schiff reaction. Whereas the method is negative for phospholipids, glycolipids are found most commonly in nervous tissue.

DERIVED LIPIDS

These contain the fatty acids that are produced by hydrolysis of the simple and compound lipids discussed previously. The fatty acids can be either saturated or unsaturated. Saturated fatty acids include palmitic and stearic acids, and these contain no double bonds in their molecule.

Unsaturated fatty acids such as oleic acid contain double bonds. Also classed in this group are the sterols, which are histochemically the more important. They are produced by the hydrolysis of ester waxes, and cholesterol is the most important histochemically.

DEMONSTRATED AND IDENTIFICATION

Lipid Material

Lipids usually occur as mixtures and because of this the individual components may not react in their characteristic manner. They usually occur as granules or droplets and in some cases bound to components within the cell. Lipid droplets can be demonstrated by staining methods as well as by extraction with organic solvents. Bound lipids are difficult to demonstrate as they are intermingled with other tissue components, and they form complex groups with other substances, e.g. with proteins to form lipoproteins. The physicochemical state of lipids in cells and tissues is variable. Cain listed the four following possibilities.

(*a*) Lipids detectable as such in living-or fixed tissues;

(*b*) Lipids present as such in living tissues but detectable only after fixation;

(*c*) Lipids not present as such in living tissues but demonstrable only after fixation (fixation unmasks certain lipids) ;

(*d*) Lipids that can be detected only after special treatment (unmasking) in addition to fixation.

Fixation

Formaldehyde

In many instances it is preferable to demonstrate lipids in unfixed tissue sections. However this is not usually possible and formaldehyde is the routine fixative of choice. Baker states that phospholipids may be lost in formol saline and suggests incorporating calcium chloride, as its buffering effect stops the loss of phospholipids.

There are grounds for a number of lipids to react chemically with formaldehyde and this should be borne in mind if considering applied lipid histochemistry. As a general rule fixation should be kept as short as possible.

Alcohol

The majority of lipids, including triglycerides and most phospholipids, are soluble in alcohol. Alcohol fixation also makes sectioning difficult and should be avoided.

Mercuric chloride

This fixative reacts with phospholipids, hydrolyzes plasmalogens, and also interferes with sectioning by making the blocks brittle and difficult to cut.

Osmium tetroxide

This fixative will react with the double bonds of unsaturated fatty acids and is involved in a number of histochemical methods for lipids.

Potassium dichromate

Potassium dichromate, like mercuric chloride, will oxidize some lipids and also binds chromium to lipid as in the Weigert Pal technique. It should be avoided as a routine fixative unless essential for the histochemical method.

Embedding

Frozen and cryostat sections are the material of choice, and are best stained free-floating for the lipid soluble dye methods. Paraffin and celloidin processing will remove the majority of lipids. Freeze drying followed by Carbowax embedding is a useful technique; Carbowax embedding of routinely fixed tissue can also be used.

IDENTIFICATION OF LIPIDS

The various lipids can be differentiated from other substances by using a number of techniques. These include physical methods, with and without dyes, and histochemical techniques for some of the complex lipids. A number of points must be considered at this stage.

The compound lipids contain some hydrophilic lipids which are water soluble; some of these can be rendered insoluble by fixation in formol calcium. The simple lipids are soluble in many organic solvents and for their demonstration only frozen sections can be used.

Many of the lipids found in tissues are bound, usually to carbohydrates or proteins, with the resulting alteration to their chemical and physical properties. A number of lipids are liquid at body temperature but may be solid at room temperature which will affect their staining properties.

Extraction

Pure hydrophobic lipids can be extracted by organic solvents and this technique is used to assist in the identification of lipids. Extraction methods usually fail to demonstrate specific tissue lipids, because of the binding of the lipid to other tissue structures. While the technique is worth attempting, the results must be viewed with caution.

The process is carried out on fresh tissue blocks placed in the extraction fluid or fluids for up to 48 hours at 60°C, or in some instances, at the boiling point of the solvent. After extraction, the block is fixed in formol saline for 24 hours, washed in tap water, frozen and suitable sections cut.

These are stained alongside unextracted tissue as a suitable control. The extraction technique can also be applied to unfixed frozen sections and with less success on fixed frozen sections. The majority of lipids are extracted by treatment with hot chloroform-methanol 2:1, cerebrosides by hot acetone, triglycerides, cholesterol and cholesterol esters by treatment with cold acetone and the majority of phospholipids are removed after exposure to hot ether.

In theory lipid extraction followed by suitable staining methods should be a control for the staining method employed, also extraction should allow for demonstration of specific lipids by their identification on the basis of their differential solubilities. The results must be viewed with caution.

General Lipid Staining Methods

The Sudan dyes have long been used to demonstrate lipids. Sudan Black and Oil Red 0 are the most recent and the most used.

The dyes are suspended in a solvent in which they are only partially soluble, and when the dye solutionn is applied to lipids (in which the dye is more soluble), transfer of the dye into the lipid occurs.

Table 14.1: Suggested lipid methods.

General lipids	Oil Red O or
Liquid and semi liquid	Sudan Black B
Triglycerides	Calcium lipase
Unsaturated lipids	Performic Acid-Schiff: OTAN
General phospholipid	Sudan Black B; OTAN
Choline-containing phospholipids	Acid haematin; OTAN
Sphingomyelin-containing phospholipids	NaOH OTAN
Plasmalogen-containing phospholipids	Mercuric Chloride-Schiff
Fatty acids	Holczingers: NaOH OTAN
Glycolipids	PAS
Cholesterol and esters	PAN; digitonin

The original dye solvent must not be capable of dissolving appreciable amounts of the lipid in the tissue section. Therefore staining occurs with these lipid-soluble dyes because the dye is more soluble in the lipid in the tissue section than it is in the original solvent.

(*a*) ***Oil Red O***

This dye is slightly soluble in organic solvents, is insoluble in water and is preferentially absorbed by lipids from solvents. The colouration obtained by using this dye is more intense than with other Sudan dyes.

(*b*) ***Sudan Black B***

Sudan Black B has a slightly different chemical structure from Oil Red O. Because of its amino groups Sudan Black B is a slightly basic dye and will combine with acidic groups in compound lipids and therefore also stain phospholipids. It has a similar solubility to Oil Red O and stains neutral lipids by preferential absorption.

(*c*) ***Nile Blue***

This method, first introduced by Lorrain Smith , has since been adapted by many others, notably by Cain. The method is usually employed to differentiate between acidic and non-acidic (neutral) lipids. Nile Blue solution contains three components.

(*a*) Oxazone salt of Nile Blue, which is dark blue in colour.

(*b*) The free base of Nile Blue oxazone, which is red in colour.

(*c*) The oxazone derivative, which is red in colour.

The first is soluble in water and alcohol, but insoluble in lipids. The second is insoluble in water but very soluble in lipids, while the last one is also insoluble in water and soluble in lipids.

When a 1 per cent solution of Nile Blue is applied to a section, neutral lipids dissolve the red component (Nile Red) in the same manner as Sudan Black and Oil Red O. Acidic lipids dissolve the oxazone and combine with the free base. The result of this combination is a salt which is lipid-soluble and dark blue in colour.

When carrying out the method, three sections are required. The first section is stained in Sudan Black B or Oil Red O to indicate areas of lipid material. The other two sections are processed through the Nile Blue method. One section is stained in 1 per cent Nile Blue at 60°C-70°C for 5 min, and the other in 0.2 per cent Nile Blue.

The weaker solution contains more of the free base. If the two Nile Blue stained sections are the same, the first is discarded. This method is not specific for any lipid group but it is useful in the process of determining whether the lipid is acidic or neutral.

(*d*) ***Osmium tetroxide***

This substance is soluble in all types of lipids; its usefulness is dependent upon its reduction to osmium dioxide by unsaturated fatty acids or choline. It will fail to react with lipids in which all the fatty acids are fully saturated, and those in which the base choline is not present. It is not used as a general lipid method, but is employed in some of the differential methods given at the end of the chapter.

Choice of Solvent for Lipid-Soluble Dyes

To obtain a satisfactory staining result, a solvent that will dissolve the dye in sufficient concentration must be used. The dye must also be more soluble in the lipid than in the solvent. If this is not the case, the final staining of the lipid material will be poor and patchy. The solvents normally used are as follows.

(*a*) 70 per cent Alcohol
(*b*) 60 per cent Isopropyl alcohol
(*c*) 60 per cent Triethylphosphate
(*d*) Propylene glycol
(*e*) 70 per cent Alcohol/absolute acetone mixture

Differential Lipid Staining Methods

Acid Haematein method

This method for phospholipids was introduced by Baker. It is based on the technique of Smith-Dietrich. It is specific only when used in conjunction with the pyridine extraction technique. Substances other than phospholipids may give a positive result, but only phospholipids are removed by pyridine extraction.

The method is carried out on a rigid time scale which includes fixation of the block. The method depends upon the formation of a dye lake by the combination of chrome ions with phospholipids. Differentiation is carried out with a borax-ferricyanide mixture. Components that contain phosphoric acid retain the dye, whereas others do not. Phospholipids are the only group of lipids to contain this acid.

O TAN method

This technique published by Adams involves the use of osmium tetroxide which is soluble in all types of lipid and is reduced by unsaturated fatty acids. The reduction is in two stages. The product of the first reduction is a colourless substance which is further reduced to produce black osmium dioxide.

The addition of sodium perchlorate can prevent this final reduction by some of the lipids. The simple lipids are not water-soluble and therefore are unaffected by the sodium perchlorate; they will continue to reduce the osmium tetroxide until black osmium dioxide is formed.

Some compound lipids such as phospholipids are partly water-soluble and the sodium perchlorate will inhibit the final reduction of osmium tetroxide by these types of lipids; the sodium perchlorate stops the reduction at the colourless intermediate substance stage.

This colourless substance is chelated with a-naphthylamine to produce an orange-red colour. The method demonstrates sites of simple lipids (triglycerides, cholesterol and its esters) and fatty acids as black, and phospholipids as orange-red.

NaOH OTAN method

This is a modification of the OTAN technique which allows the separation of lecithin from sphingomyelin. The choline-containing phospholipids such as lecithin have ester linkages which will be broken by alkaline hydrolysis, the result being no osmium staining.

The linkage of the fatty acid to sphingosine is not affected and is in consequence stained black. The plasmalogen linkage also resists treatment with sodium hydroxide and will also react with osmium tetroxide but it can be blocked by treatment with mercuric chloride.

The Plasmal reaction

The acetal phosphatides (plasmalogens) are phospholipids that contain one molecule of an aldehyde of a fatty acid which combines with the glycerol group, the linkage between the two components being of the acetal type. Feulgen and Voit introduced a method to demonstrate these lipids.

The acetal linkage is split by exposure to mercuric chloride. The aldehydes are released, and these can then be demonstrated by the Periodic Acid-Schiff method. Terner and Hayes described this method in great detail, but there is doubt regarding its specificity.

Their general conclusion was that the plasmalogens were being demonstrated. For further details, readers are referred to the original paper and to Pearse.

Performic Acid-Schiff method

This method depends upon the oxidation of unsaturated bonds (ethylenic bonds) to produce free aldehyde. This is demonstrated by the Schiff's reagent. The oxidation can be carried out using performic acid or peracetic acid.

Performic acid is used in preference to peracetic acid. The method is not specific for unsaturated lipids, as keratin and DNA will also react. If blocking is carried out using bromine water, staining of the unsaturated lipid is prevented. These two techniques should always be carried out simultaneously when unsaturated lipids are being demonstrated.

PAS reaction for glycolipids

The reaction for glycolipids depends upon their content of D-galactose. This sugar is PAS-positive. All the aldehydes in the section are first blocked, using Method 31. This technique will distinguish between glycolipids and other lipid groups, but not between the two members of the glycolipid group: cerebrosides and gangliosides.

The Perchloric Acid-Naphthoquinone method

The standard method for the demonstration of cholesterol was, at one time, the Schultz method. This has been replaced by a more sensitive method introduced by Adams. His technique is known as the Perchloric Acid-Naphthoquinone method (PAN).

The basis of the reaction is that perchloric acid first forms an insoluble perchlorate with cholesterol. This is then converted to cholesta-3 :5-diene by the elimination of water, and this subsequently reacts with 1:2-naphthoquinone-4sulphonic acid to form a dark blue pigment. The pigment is stable for a few hours, but the section should be examined immediately.

Digitonin method

This technique is used to distinguish between cholesterol and cholesterol esters. Windaus introduced the technique which requires two formalin-fixed frozen sections. The first section is stained by the Oil Red 0 method for lipid material.

The second section is placed in a solution of digitonin. Free cholesterol forms a complex with the digitonin, and characteristic birefringent crystals are precipitated from this solution. The section is examined under a polarizing microscope.

Birefringent needle-like crystals indicate free cholesterol. The control section stained in Oil Red O will also exhibit birefringence of free cholesterol, whereas esters will be stained by Oil Red O and will not be birefringent.

The Copper-Rubeanic Acid method

Methods suitable for the demonstration of free fatty acids are restricted to modifications of the technique introduced by Fischler. Holczinger reviewed a number of these methods and recommended the Copper-Rubeanic Acid method as the best available technique.

Sections are exposed to copper acetate for 2-5 hours, followed by treatment in ethylenediamine tetra-acetic acid (EDTA) to remove non-specific absorbed copper. Dilute rubeanic acid is applied to the sections and copper soaps are produced indicating the presence of fatty acid deposits.

Identifying a Lipid

The process of identifying a lipid is complicated, and may involve using four or more staining methods as well as extraction techniques. When lipid material is suspected, two sections of fixed frozen material are used. The first is stained by a lipid-soluble dye method, which will reveal the presence of any lipid material.

The second section is looked at unstained. In this state, refractile droplets indicate lipid material. The section is then viewed by a polarizing microscope, for if the lipid material is cholesterol or cholesterol esters these can be detected by their birefringence.

Once lipid material has been confirmed, it is necessary to test whether it is acidic or non-acidic (neutral). This is done by applying the Nile Blue technique Acidic lipids are stained blue, and the non-acidic (neutral) droplets, red. If the lipid material is acidic, this indicates that it is either a phospholipid or a fatty acid.

If it is non-acidic, it may be one of the following: an unsaturated lipid, a glycolipid, cholesterol and cholesterol esters, or a simple lipid. The procedure outlined above should be undertaken in order to complete identification. The extraction of lipids plays an important role in their identification, and should be used in conjunction with the various staining methods.

Method 14.1

Lipid material : Oil Red O

Reagents required

(*a*) Oil Red O

(*b*) Triethyl phosphate

(*c*) Distilled water

Preparation of staining solution

Oil Red O	1 g
Triethyl phosphate	60 ml
Distilled water	40 ml

The distilled water is added to the triethyl phosphate, the dye is then added, the mixture is heated to 100°C for 5 min and is stirred constantly. The mixture is filtered when hot and again when cool. This solution will keep as a stock solution but must be filtered before use.

Sections

Formalin-fixed frozen sections, free-floating

Cryostat post-fixed, preferably free-floating

Method

(*a*) Wash sections in distilled water

(*b*) Place sections in 60 per cent triethyl phosphate

(*c*) Stain sections in Oil Red O solution at 20°C for 15 min

(*d*) Wash sections in 60 per cent triethyl phosphate for 30 sec

(*e*) Wash sections in distilled water

(*f*) Stain sections in haematoxylin for 1 min

(*g*) Wash sections in tap water for 5 min
(*h*) Mount in glycerin jelly

Results

Lipid material: red
Nuclei : blue

Remarks

This method can be employed at 37°C or 60°C if required.

Method 14.2

Lipid material: Sudan Black B

Reagents required

(*a*) Sudan Black B
(*b*) Triethyl phosphate
(*c*) Distilled water

Preparation of staining solution

Sudan Black B	1 g
Triethyl phosphate	60 ml
Distilled water	40 ml

The distilled water is added to the triethyl phosphate, the stain is added to this solution and the mixture is heated to 100°C for 5 min and stirred constantly. The mixture is filtered when hot and once again immediately before use. The solution will keep well as a stock solution, but must be filtered each time it is used.

Sections

Formalin-fixed frozen sections, free-floating
Cryostat post-fixed, preferably free-floating

Method

(*a*) Wash sections in distilled water
(*b*) Place sections in 60 per cent triethyl phosphate
(*c*) Stain sections in Sudan Black B solution at 20°C for 10 min
(*d*) Place sections in 60 per cent triethyl phosphate for 30 sec
(*e*) Wash in distilled water
(*f*) Stain in Mayer's Carmalum for 3 min
(*g*) Wash in distilled water
(*h*) Mount in glycerin jelly

Results

Lipid material, including phospholipids: black

Nuclei: red

Remarks

This technique using triethyl phosphate can be carried out at 37°C or 60°C as required.

Method 14.3

Acidic lipids: Nile Blue

Reagents required

(*a*) Nile Blue
(*b*) Acetic acid
(*c*) Distilled water

Preparation of solutions

(*a*) Nile Blue 1 per cent

Nile Blue	500 mg
Distilled water	50 ml

(*b*) Nile Blue 0·02 per cent

Nile Blue	10 mg
Distilled water	50 ml

(*c*) D ferentiator

Acetic acid, conc.	0·5 ml
Distilled	50 ml

Sections

Formalin-fixed frozen sections
Cryostat pre-fixed
Cryostat post-fixed

Method

It is necessary to stain one section by the Oil Red O or Sudan Black methods. This slide is numbered 3. Two sections are required for the following technique.

(*a*) Bring both sections to water
(*b*) Stain sections 1 and 2 in 1 per cent Nile Blue for 5m in at 60°C (solution 1)
(*c*) Differentiate at 60°C in solution (*c*) for 30 s
(*d*) Wash in tap water
(*e*) Mount section 1 in glycerin jelly
(*f*) Place section 2 in solution (*b*) (0·02 per cent Nile Blue) for 5 min at 60°C

(*g*) Wash in tap water
(*h*) Differentiate section 2 in solution (*c*) at 60°C for 30 s
(*i*) Wash in tap water
(*j*) Mount in glycerin jelly

Results

Section 1. Any blue staining that can be compared with a positive section 3 is taken to be acidic lipid material.

Section 2. Any red staining that can be compared with a positive result in section 3 is taken to be a non-acidic lipid.

Section 3. Control for above.

Remarks

This method has caused much controversy in the literature, and its specificity is doubtful, it is however well worth applying when trying to identify an unknown lipid.

Method 14.4

Phospholipids: Acid Haematein method

Reagents required

(*a*) Formalin
(*b*) Calcium chloride
(*c*) Potassium dichromate
(*d*) Haematoxylin
(*e*) Sodium iodate
(*f*) Glacial acetic acid
(*g*) Sodium tetraborate (borax)
(*h*) Potassium ferricyanide
(*i*) Distilled water

Preparation of solutions

(*a*) Fixative

Formalin	10 ml
Calcium chloride (anhydrous)	1 g
Distilled water	90 ml

(*b*) Post-chroming solution

Potassium dichromate	5 g
Calcium chloride	1 g
Distilled water	100 ml

(*c*) Acid Haematein solution

Haematein	50 mg

1 per cent Sodium iodate	1 ml
Distilled water	49 ml

Heat the solution to boiling point, allow to cool and add 1 ml of glacial acetic acid.

(*d*) Differentiator

Potassium ferricyanide	250 mg
Sodium tetraborate (borax)	250 mg
Distilled water	100 ml

Sections

For this method unfixed pieces of tissue are recommended

Method

(*a*) Place blocks in fixative, solution (*a*) at 22°C, for 6-12 h

(*b*) Transfer tissue (no washing) to post-chroming solution (*b*) at 22°C for 18 h

(*c*) Transfer tissue to fresh post-chroming solution (*b*) at 60°C, for 24 h

(*d*) Wash in running tap water for 6 h

(*e*) Cut frozen sections 10 μm thick

(*f*) Place sections in post-chroming solution (*b*) at 37°C, for 1 h

(*g*) Wash well in distilled water for 5 min

(*h*) Stain in acid haematein, solution (*c*), at 60°C, for 5 h

(*i*) Rinse well in distilled water

(*j*) Transfer sections to differentiating solution (*d*) at 37°C for 18h

(*k*) Wash in tap water for 10 min

(*l*) Mount in glycerin jelly

Results

Phospholipids: dark blue

Other material may be blue (*see Remarks*)

Remarks

This method is only specific when a control is submitted to the pyridine extraction method (*see* Method 14.5).

Method 14.5

Phospholipids : Pyridine Extraction

Reagents required

(*a*) Picric acid

(*b*) Formalin
(*c*) Glacial acetic acid
(*d*) Distilled water
(*e*) 70 per cent Alcohol
(*f*) 50 per cent Alcohol
(*g*) Pyridine
(*h*) Distilled water

Preparation of solutions

Fixative (*Bouin's*)

Saturated aqueous picric acid	50 ml
Formaldehyde	10 ml
Glacial acetic acid	5 ml
Distilled water	35 ml

Sections

To act as ,a control for Method 14.4 a comparable block must be used.

Method

(*a*) Fix the block in the dilute Bouin's fixative 20 h
(*b*) Wash in 70 per cent alcohol 1 h
(*c*) Wash in 50 per cent alcohol 30 min
(*d*) Wash in running tap water 30 min
(*e*) Place in pyridine at 22°C 1 h
(*f*) Transfer block to fresh pyridine at 22°C 1 h
(*g*) Transfer block to fresh pyridine at 60°C 24 h
(*h*) Wash in running tap water 2 h
(*i*) Transfer to stage (*b*) of Method 14.4

Results

After the haematein method has been applied to the blocks processed by the above technique, phospholipids are negative.

Method 14.6

OTAN method

Reagents required

(*a*) Osmium tetroxide
(*b*) Potassium perchlorate
(*c*) α-naphthylamine

Preparation of solutions

(*a*) Osmium tetroxide solution

1 per cent osmium tetroxide	1 part
1 per cent potassium perchlorate	3 parts

(*b*) α-naphthylamine solution. Saturated solution in warm distilled water, β-naphthylamine may be a contaminant and is a carcinogen.

Sections

Frozen sections (formol-calcium-fixed)

Method

(*a*) Treat free-floating sections with osmium solution in suitable container filled with fluid and tightly stoppered for 18 h
(*b*) Wash in distilled water for 10 min
(*c*) Pick up on slides
(*d*) Treat with a-naphthylamine solution at 37°C for 20 min
(*e*) Wash sections in distilled water for 5 min
(*f*) Mount in glycerin jelly

Results

Phospholipids: orange-red
Cholesterol esters : black
Triglycerides: black

Remarks

Sections may be counterstained after stage (*e*) if required.

Method 14.7

Sodium Hydroxide OTAN method

Reagents required

(*a*) Osmium tetroxide
(*b*) Potassium perchlorate
(*c*) α-naphthylamine
(*d*) Sodium hydroxide

Preparation of solutions

(*a*) Osmium tetroxide (see Method 14.6)
(*b*) α-napthylamine (see Method 14.6)
(*c*) Sodium hydroxide

Sodium hydroxide	8 g
Distilled water	100 ml

Sections

Frozen sections (formol-calcium-fixed)

Method

(*a*) Treat free-floating sections with sodium hydroxide solution at 37°C for 1 h

(*b*) Wash in distilled water

(*c*) Treat sections in 1 per cent acetic acid

(*d*) Continue from step (*a*) in previous method

Results

Sphingomyelin : orange-red

Other alkali resistant lipids will stain black-alkali labile lipids will be destroyed.

Method 14.8

Acetal phosphatides: the Plasmal reaction

Reagents required

(*a*) Sodium chloride

(*b*) Mercuric chloride

(*c*) Hydrochloric acid

(*d*) Schiff's reagent

(*e*) Sodium metabisulphite

Preparation of solutions

(*a*) Sodium chloride solution

Sodium chloride	900 mg
Distilled water	100 ml

(*b*) Mercuric chloride solution

Mercuric chloride	1 g
Distilled water	100 ml

(*c*) Schiff's reagent

(*d*) Sulphurous acid rinse

Sodium bisulphite	500 mg
Distilled water	99.5 ml
Conc. hydrochloric acid	0.5 ml

Sections

Frozen formalin-fixed

Frozen unfixed

Cryostat post-fixed

Cryostat pre-fixed

Method

With this technique it is necessary to have a control section that is not placed in solution (*b*), the mercuric chloride.

(*a*) If section is unfixed, place in three changes of solution (*a*) otherwise

(*b*) Wash fixed sections in distilled water.

All sections:

(*c*)	Place section in solution (*b*) mercuric chloride	7 min
(*d*)	Transfer sections to Schiff's reagent, solution (*c*)	10 min
(*e*)	Transfer sections to sulphurous acid rinse 1	2 min
(*f*)	Transfer sections to sulphurous acid rinse 2	2 min
(*g*)	Transfer sections to sulphurous acid rinse 3	2 min
(*h*)	Wash in tap water	2 min
(*i*)	Counterstain in 2 per cent Methyl Green	3 min
(*j*)	Wash in running tap water	
(*k*)	Mount in glycerin jelly	

Results

Acetal phosphatides: Schiff-positive, reddish-pink. The control section should be negative.

Remarks

It is possible that aldehydes are present in the section, in which case a false positive may be obtained.

Method 14.9

Unsaturated lipids: Perforinic Acid—Schiff

Reagents required

(*a*) Formic acid

(*b*) Hydrogen peroxide

(*c*) Sulphuric acid

(*d*) Schiff's reagent

Solutions

(*a*) Performic acid

90 per cent Formic acid,	40 ml
100 vol. or 30 per cent Hydrogen peroxide	4 ml

Sulphuric acid (conc.) 0·5 ml

Allow to stand for 1 h before use

(*b*) Schiff's reagent

Sections

Cryostat post-fixed Cryostat pre-fixed Formalin-fixed frozen sections Freeze dried paraffin sections

Method

(*a*) Bring all sections down to tap water
(*b*) Treat sections with solution (*a*) for 30 min
(*c*) Wash in tap water for 15 min
(*d*) Immerse in Schiff's reagent (solution 2) for 40 min
(*e*) Wash in running tap water
(*f*) Dehydrate through alcohols to xylene
(*g*) Wash and mount in DPX

Results

Unsaturated lipids including cerebrosides : red

Remarks

(*a*) The performic acid solution must be allowed to stand for 1 h before use, and can only be used on the day it is prepared.
(*b*) For control method see bromination technique (Method 14.10).

Method 14.10

Unsaturated lipids: Bromination

Reagents required

(*a*) Carbon tetrachloride
(*b*) Bromine water
(*c*) Sodium metabisulphite
(*d*) Absolute alcohol

(a) ***Frozen Sections***

Solutions required

(*a*) 2.5 per cent Bromine water

Bromine water	1 ml
Distilled water	39 ml

(*b*) 0.5 per cent Sodium metabisulphite

Sodium metabisulphite	500 mg
Distilled water	50 ml

Method

(*a*) Attach sections to slides and allow to dry
(*b*) Immerse sections in 2.5 per cent bromine water 1-6 h
(*c*) Wash in tap water
(*d*) Treat sections with 0.5 per cent sodium metabisulphite 2 min
(*e*) Wash in running tap water
(*f*) Apply method

(b) *Paraffin Sections*

Solution

Carbon tetrachloride	39 ml
Bromine water	1 ml

Method

(*a*) Dewax sections in xylol
(*b*) Immerse sections in carbon tetrachloride, 2 changes 4 min
(*c*) Immerse sections in bromine solution 1 h
(*d*) Immerse sections in carbon tetrachloride, 2 changes 4 min
(*e*) Rinse sections in 90 per cent alcohol 1 min
(*f*) Rinse sections in 60 per cent alcohol 1 min
(*g*) Rinse sections in water
(*h*) Apply method

Method 14.11

Glycolipids: PAS reaction

Reagents required

(*a*) Periodic acid
(*b*) Schiff's reagent
(*c*) Distilled water

Preparation of solutions

(*a*) Periodic acid

Periodic acid	500 mg
Distilled water	100 ml

(*b*) Schiff's reagent

Sections

Cryostat post-fixed

Cryostat pre-fixed
Formalin-fixed frozen sections
Freeze dried paraffin sections

Method

(*a*)	Bring sections down to water	
(*b*)	Transfer sections to solution (*b*)	5 min
(*c*)	Wash in tap water	3 min
(*d*)	Place in Schiff's reagent	20 min
(*e*)	Wash in tap water	20 min
(*f*)	Counterstain in haemalum	5 min
(*g*)	Wash in tap water	
(*h*)	Differentiate in 1 per cent acid alcohol	5 s
(*i*)	Wash in tap water	
(*j*)	Mount in glycerin jelly	

Results

Glycolipids, mucins, etc.: red
Nuclei: blue

Remarks

It is necessary to use the following controls when applying this method for glycolipids.

(*a*) Oil Red O or Sudan Black to confirm site of possible glycolipid.

(*b*) Aldehydes to be blocked.

Note: If formalin-fixed frozen sections are used, they must be well washed in several changes of distilled water to remove any free aldehydes. If cryostat sections are used, alcohol should be used as a fixative.

Method 14.11

Cholesterol and related substances:
Perchloric Acid Naphthoquinone reaction

Reagents required

(*a*) 1:2-Naphthoquinone-4-sulphonic acid
(*b*) Ethanol
(*c*) Perchloric acid
(*d*) Formaldehyde
(*e*) Distilled water

Preparation of solution

1:2-Naphthoquinone-4-sulphonic acid	12 mg

Ethanol	6 ml
60 per cent Perchloric acid	3 ml
Conc. formaldehyde	0.3 ml
Distilled water	2.7 ml

The ethanol-perchloric acid-formaldehyde-water solution is prepared first and the reagent dissolved in it.

Sections

Formol saline-fixed, frozen sections, free-floating Formol calcium-fixed, frozen sections, free-floating

Suitable sections for controls

Adrenal gland

Method

(*a*) Cut frozen sections and float into formalin. Leave for 7 days
(*b*) Mount sections on slides and dry at room temperature
(*c*) Soak sections in reagent
(*d*) Heat sections in reagent to 60-70°C for 10 min
(*e*) Mount section in 60 per cent perchloric acid

Results

Cholesterol and its esters: dark blue

Remarks

(*a*) The dark blue colour is stable for a few hours.
(*b*) During the heating, the section should change colour from red to dark blue.

Method 14.13

Free cholesterol: Digitonin method

Reagents required

(*a*) Ethyl alcohol
(*b*) Digitonin
(*c*) Distilled water

Preparation of solutions

(*a*) Ethyl alcohol 50 per cent

Ethyl alcohol	100 ml
Distilled water	100 ml

(*b*) Digitonin solution

Digitonin	500 mg
Solution (*a*)	100 ml

Sections

Formalin-fixed frozen sections free-floating. A control section is stained by the Oil Red 0 method.

Method

(*a*) Incubate sections for 3 h in solution
(*b*) at room temperature (*b*) Rinse sections in solution (*a*)
(*c*) Float sections onto slides
(*d*) Mount in glycerin jelly

Results

Digitonin section: free cholesterol, birefringent.

Oil Red O section: free cholesterol, birefringent. Cholesterol ester, stained by Oil Red O.

Method 14.14

Fatty acids: Copper-Rubeanic Acid method

Reagents required

(*a*) Copper acetate
(*b*) Rubeanic acid
(*c*) Ethanol
(*d*) Distilled water
(*e*) Ethylenediamine tetra-acetic acid (Disodium), EDTA

Preparation of solutions

(*a*) 0.005 per cent Copper acetate

Copper acetate	5 mg
Distilled water	100 ml

(*b*) 0.1 per cent EDTA

Ethylenediamine tetra-acetic acid	50 mg
Distilled water	50 ml

(*c*) 0.1 per cent Rubeanic acid

Rubeanic acid	50 mg
Absolute alcohol	35 ml
Distilled water	15 ml

Dissolve the rubeanic acid in the absolute alcohol by warming slightly then add the distilled water.

Sections

Cryostat unfixed
Cryostat pre-fixed

Formalin-fixed frozen sections

Method

(*a*)	Place sections in copper acetate solution	3—5 h
(*b*)	Wash sections in EDTA solution	10 sec
(*c*)	Wash sections again in EDTA solution	10 sec
(*d*)	Wash sections in distilled water	10 min
(*e*)	Immerse sections in rubeanic acid solution	30 min
(*f*)	Wash sections in 70 per cent alcohol	3 min
(*g*)	Wash sections in running tap water	
(*h*)	Mount sections in glycerin jelly or dehydrate through graded alcohols and mount in DPX	

Results

Fatty acids: greenish black

15

NUCLEIC ACIDS

The nucleic acids *deoxyribonucleic acid* (DNA) and *ribosenucleic acid* (RNA), are found in all animal and plant tissues, and are usually combined with basic proteins to form so-called *nucleoprotein*. DNA is mainly found in the nucleus, and RNA in the *cytoplasm*.

Hydrolysis of nucleic acids yields the following components : (i) *phosphate* groups, (ii) five-carbon sugars, and (iii) *nitrogenous bases*, *purines* and *pyrimidines*. The structure is in the form of a ladder, the side-arms of which are composed of alternate sugar and phosphate groups.

The transverse struts are formed by the nitrogenous bases linked to each other and to the sugar groups of the side arms of the ladder. The nitrogenous bases are of four types, two of which are purines and two are *pyrimidines*; in the cross-struts a *purine* is always linked to a *pyrimidine*. The arrangement can be diagrammatically represented thus:

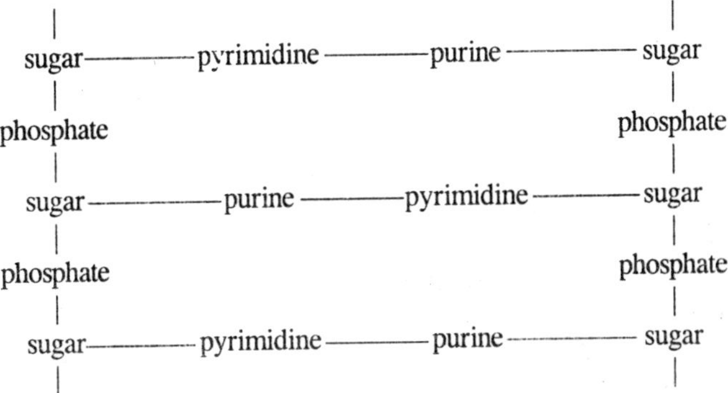

The final structure is this ladder arrangement twisted into a spiral. Both DNA and RNA have this basic structure, but differ in two important respects. These are:

(*i*) *Sugar content*. In DNA the 5-carbon sugar is deoxyribose, but in RNA it is ribose.

(*ii*) *Purine and pyrimidine content*. In DNA the purines are adenine and guanine, and the pyrimidines are thymine and cytosine. In RNA the purines are the same as in DNA, but the pyrimidines are uracil and cytosine.

DEMONSTRATION OF DNA AND RNA

The large number of phosphate radicals in the nucleic acid molecule ensure that nucleic acids will take up basic dyes such as haematoxylin. Thus the nucleus, with a high DNA content, will stain blue-black with haematoxylin.

The cytoplasm of most cells do not contain enough RNA for the haematoxyphilia to overcome the eosinophilia of the large amount of protein present in the rest of the cytoplasm.

In some cells however, e.g. plasma cells, there is such a large amount of RNA present that the haematoxyphilia almost overcomes the red staining of the cytoplasm, and the final staining colour of the cytoplasm is purplish.

The following histochemical methods are used for the identification of DNA and RNA.

Feulgen method for	DNA	Method 15.1
Methyl Green—Pyronin methods for	RNA, DNA	Method 15.2, 15.3
Gallocyanin-Chrome Alum method for	RNA, DNA	Method 15.4
NAH—Feulgen method for	DNA	Method 15.5
Acridine Orange method for	RNA, DNA	Method 15.6
Deoxyribonuclease extraction for	DNA	Method 15.7
Ribonuclease extraction for	RNA	Method 15.8

Feulgen Nucleal Method

This technique, first introduced byFeulgenand Rossenbeck, is considered to be specific for deoxyribonucleic acid (DNA) since only nucleic acids containing the sugar deoxyribose will show the reaction. It is the most reliable method available for the demonstration of DNA. The reaction involves mild hydrolysis of deoxyribose sugars with warm hydrochloric acid; ribose sugar is not hydrolysed by NHCl and therefore RNA does not show the Feulgen reaction.

Hydrolysis releases free aldehydes and Schiff's reagent is then applied, to produce a coloured compound indicating the presence of DNA. Two most important factors contributing to the outcome of the Feulgen reaction are (a) the means of fixation of the tissue and (b) the duration of hydrolysis in dilute acid.

The hydrolysis time, which varies with the type of fixative used, is critical. Each fixative has its own optimum. By prolonging the hydrolysis, a stronger result is produced until the optimal time is reached. If this is exceeded, the intensity of staining becomes weaker and may eventually disappear altogether. Excessive hydrolysis results in total extraction of DNA from the section.

Bauer published a paper dealing with optimal hydrolysis times for different fixatives. These figures are reproduced in Table elsewhere in this chpater, together with some additional data. Formol saline buffered to pH 7.2 or Carnoy's are recommended fixatives.

Bouin's fixative should not be used, as this causes over-hydrolysis of the nucleic acid during fixation. Control sections are incubated in deoxyribonuclease before hydrolysis after which the technique will prove negative.

Methyl Green-Pyronin

The Methyl Green-Pyronin technique is able to demonstrate both DNA and RNA. The method was first published by Pappenheim and modified by Unna. It is often referred to as the Unna Pappenheim Stain. Several attempts have since been made to render the method more specific.

Methyl Green is an impure dye containing Methyl Violet. To obtain satisfactory results with this method it is necessary to wash the Methyl Green (2 per cent aqueous solution) several times with equal parts of chloroform.

Six or more washings may be required before all the violet is removed. When treated in this way, Methyl Green seems to be specific for DNA at a slightly acid pH. Methyl Green is a basic dye, and its affinity for DNA is not fully understood.

However, Kurnick has suggested that binding of Methyl Green to DNA involves two sites, two amino groups on the dye combining with two phosphoric acid groups of the DNA. If the dye is used at a slightly acid pH, protein material does not stain.

Pyronin Y, used in the method, is less specific and must be used under carefully regulated conditions. Precisely-controlled staining times must be used, and the pH of the staining solution is critical. When using pyronin solutions, it is always advisable to use a control section which has been subjected to ribonuclease digestion.

Pyronin Y obtained commercially varies, and certain batches fail to give satisfactory results. Because of this, it is necessary to compare each batch on its arrival against a known good batch of dye. The chemical basis of the method is not fully understood. It is discussed in considerable detail by Pearse.

Gallocyanin-Chrome Alum Method

This method, first introduced by Einarson for Nissl granules, was later adapted to demonstrate nucleic acids. The method does not distinguish between DNA and RNA, but with suitable extraction methods it is possible to identify one or the other when required. The method is simple to perform and is a progressive stain.

The mechanism is due to the combination at an acid pH of the phosphoric acid residue of the nucleic acid with gallocyanin. The reaction must be carried out at a pH of 1 because at higher values (pH 2, or above), other tissue components will stain.

Feulgen-Naphthoic Acid-Hydrazide Reaction

This method is recommended as a control method for the Feulgen nucleal technique, because material giving a possible positive Schiff result (i.e., some lipoproteins) will not be stained by this method. The sections are hydrolysed in the manner described earlier, using normal hydrochloric acid.

The free aldehydes produced on hydrolysis combine with 2-hydroxy-3-naphthoic acid hydrazide. This is then coupled to Fast Blue B. producing a purplish blue colour at the site of coupling. The localisation is identical to that shown by the true Feulgen reaction. Danielli originally used 2 :4-dinitrophenylhydrazine, but better results are obtained using 2-hydroxy-3-naphthoic acid hydrazide.

Acridine Orange

This fluorescent technique was introduced by Von Bertalanffy and Bickis and is widely used in exfoliative cytology. The technique is for DNA and' RNA. Nuclear DNA fluoresces apple green while RNA fluoresces red, as do the mucins. The successful application of this method depends upon the concentration of Acridine Orange, the pH of the staining solution and the fixative applied to the fresh frozen sections or to the smears. In the recommended method, the strength of Acridine Orange is 0.1 per cent in phosphate buffer, pH 6-0.

If the Acridine Orange is too concentrated the red colouration masks the green. Formalin and Bouin fixation must be avoided as these both prevent differential staining. Acetic ethanol or 70 per cent alcohol are recommended for use.

DIGESTION METHODS

The two enzyme techniques for the digestion of nucleic acids are invaluable as controls for the methods discussed. Deoxyribonuclease is specific for the removal of DNA in sections, while not affecting the RNA content. When ribonuclease is applied to tissue sections, all the RNA is removed whilst the DNA is unaltered.

The digestion methods were introduced by Brachet as controls for the Methyl Pyronin method. Great care must be taken that the enzymes are of a high purity, for impure enzymes will remove all nucleic acids. Both ribosenuclease and deoxyribosenuclease are expensive reagents.

A useful digestion method which is not so specific as the enzyme digestion methods, is that which uses perchloric acid. It will either remove RNA alone or both RNA and DNA according to the conditions under which the extractions are controlled.

Table 15.1: Hydrolysis in N-HCl at 60°C

Fixative	*Time (min)*
Bouin	not recommended
Carnoy	6
Flemming's	8
Formalin	8-10
Newcomer's	20
Susa	18
Zenker	5

Method 15.1

DNA: Feulgen nucleal reaction

Reagents required

(*a*) Hydrochloric acid

(*b*) Schiff's reagent

(*c*) Potassium metabisulphite

(*d*) Light green, 1 per cent aqueous

Preparation of solutions

(*a*) N-HCl

Hydrochloric acid, (conc.)	8.5 ml
Distilled water	91.5 ml

(*b*) SchiJ's reagent

(*c*) Bisulphite solution

10 per cent Potassium metabisulphite	5 ml
N-Hydrochloric acid	5 ml
Distilled water	90 ml

Sections

All types

Suitable control sections

Pancreas

Method

(*a*)	Bring all sections to water	
(*b*)	Rinse sections in N-HCI at room temperature	1 min
(*c*)	Place sections in N-HCl at 60°C	
(*d*)	Rinse sections in N-HCl at room temperature	1 min
(*e*)	Transfer sections to Schiff's reagent	45 min
(*f*)	Rinse sections in bisulphite, solution (*c*)	2 min
(*g*)	Repeat wash in bisulphite, solution (*c*)	2 min
(*h*)	Repeat wash in bisulphite, solution (*c*)	2 min
(*i*)	Rinse well in distilled water	
(*j*)	Counterstain if required in 1 per cent Light Green	2 min
(*k*)	Wash in water	
(*l*)	Dehydrate through graded alcohols to xylene and mount	

Results

DNA: red-purple

Cytoplasm: green

Remarks

(*a*) The hydrolysis time is important (*see* text), and the correct time for the fixative must be used.

(*b*) The N-HCl should be preheated to 60°C.

Method 15.2

RNA, DNA: Methyl Green-Pyronin method

Reagents required

(*a*) Methyl Green

(*b*) Chloroform
(*c*) Pyronin Y
(*d*) 0.1 M Acetate buffer, pH 4.8

Preparation of solutions

(*a*) Methyl Green

Methyl green	2 g
Distilled water	100 ml

Dissolve the Methyl Green in the distilled water by stirring well.

Pour the solution into a separating funnel. Add 100 ml chloroform and shake well; pour off contaminated chloroform and repeat until no more violet is extracted (about 6-8 washes).

(*b*) Pyronin Y

Pyronin Y	2 g
Distilled water	100 ml

(*c*) Staining solution

Methyl Green	7.5 ml
Pyronin Y	12.5 ml
Acetate buffer, pH 4.8	30.0 ml

Sections

All types, freeze dried recommended

Suitable control section

Pancreas

Method

(*a*) Bring all sections to water
(*b*) Stain in Methyl Green—Pyronin solution, 4—10 min
(*c*) Blot dry
(*d*) Rinse rapidly in absolute acetone
(*e*) Rinse rapidly in 10 per cent acetone in xylene
(*f*) Rinse rapidly in 50 per cent acetone in xylene
(*g*) Rinse in xylene
(*h*) Place sections in fresh xylene and mount in DPX

Results

DNA : green
RNA: red

Remarks

(*a*) This method can be unreliable, the main causes being that some samples of Pyronin Y will not work satisfactorily, or that the Methyl Green is not pure enough.

(*b*) The dehydration through acetone and xylene should be rapid.

(*c*) Triethyl phosphate may be used for dehydration instead of acetone.

(*d*) Methyl Green alone may be used in the manner described above to demonstrate DNA.

Method 15.3

Methyl Green-Pyronin method for RNA, DNA

Reagents required

(*a*) 2 per cent Methyl Green. See previous method

(*b*) 5 per cent Pyronin Y

(*c*) Acetate buffer pH 4.8.

Solution A

2 per cent Methyl Green (chloroform-washed) 10 ml

5 per cent Pyronin Y 17.5 ml Distilled water 250 ml

Solution B

Acetate buffer pH 4.8

Working solution

Solution A	25 ml
Solution B	25 ml

Method

(*a*) Bring sections to water

(*b*) Rinse in distilled water and blot dry

(*c*) Stain in working solution 20—30 min

(*d*) Rinse rapidly in distilled water and blot dry

(*e*) Dehydrate in acetone

(*f*) Rinse in acetone xylene 50:50, clear in xylene

(*g*) Mount in DPX

Results

DNA: green—bluish green

RNA : red

Remarks

(*a*) Fixation should be in neutral fixatives

Method 15.4

RNA, DNA: Gallocyanin-Chrome Alum method

Reagents required

(*a*) Chrome alum

(*b*) Gallocyanin

(*c*) Distilled water

Preparation of solution

Chrome alum	5 g
Distilled water	100 ml
Gallocyanin	150 mg

The chrome alum is dissolved in the distilled water, the gallocyanin added and the solution slowly heated until it boils. It is allowed to boil for 5 min. When the solution has cooled to room temperature, the volume is adjusted to 100 ml. The solution is filtered before use.

Sections

All types

Suitable control sections

Pancreas

Method

(*a*) Bring sections down to water

(*b*) Stain in gallocyanin-chrome alum solution 18-48 h

(*c*) Wash in tap water

(*d*) Dehydrate through graded alcohols and mount in DPX

Results

RNA, DNA: blue

Remarks

This method does not distinguish between RNA and DNA but is specific for nucleic acids.

Method 15.5

DNA: Naphthoic Acid Hydrazine-Feulgen method

Reagents required

(*a*) 2-Hydroxy-3-naphthoic acid hydrazine

(*b*) Absolute alcohol

(*c*) Acetic acid

(*d*) Fast Blue B

(*e*) Veronal acetate buffer, pH 7.4

(*f*) N-Hydrochloric acid

Preparation of solutions

(*a*) N-Hydrochloric acid

Hydrochloric acid, (conc.)	8.5 ml
Distilled water	91.5 ml

(*b*) NAH solution

2-Hydroxy-3-naphthoic acid hydrazide	50 mg
Absolute alcohol	47 ml
Acetic acid (conc.)	3 ml

(*c*) Fast Blue B solution

Fast Blue B	50 mg
Veronal acetate buffer, pH 7·4	50 ml

This solution must be freshly prepared.

Sections

All types

Suitable control sections

Pancreas

Method

(*a*) Bring all sections to water (*b*) Rinse briefly in N-HCl

(*c*) Place sections in N-HCl at 60°C

(*d*) Rinse sections in N-HCl at room temperature	1 min
(*e*) Rinse sections in distilled water	1 min
(*f*) Rinse sections in 50 per cent alcohol	1 min
(*g*) Place sections in NAH solution at room temperature	3-6 h
(*h*) Rinse sections in 50 per cent alcohol	10 min
(*i*) Rinse sections in 50 per cent alcohol	10 min
(*j*) Rinse sections in 50 per cent alcohol	10 min
(*k*) Rinse sections in distilled water 1 min	
(*l*) Place sections in fresh Fast Blue B solution	3 min

(*m*) Dehydrate through graded alcohols to xylene and mount in DPX

Result

DNA: blue to bluish purple

Protein material: possibly purplish red

Method 15.6

RNA, DNA: Acridine Orange

Reagents required

(*a*) Acridine Orange
(*b*) 0.2 M Phosphate buffer
(*c*) Calcium chloride
(*d*) Distilled water
(*e*) Acetic acid

Preparation of solutions

Acridine Orange solution

Acridine Orange	50 mg
Distilled water	40 ml

The pH of the solution is adjusted to 6·0 with phosphate buffer. The volume is then made up to 50 ml.

(*a*) Phosphate buffer, pH 6.0

(*b*) Calcium chloride

Calcium chloride	11 g
Distilled water	50 ml

Sections

Frozen sections
Cryostat sections
Paraffin sections
Freeze dried sections
Smears

Suitable control sections

Pancreas

Staining method

(*a*)	Bring all sections to distilled water	
(*b*)	Rinse briefly in 1 per cent acetic acid	15 s
(*c*)	Rinse sections in distilled water	15 s
(*d*)	Stain in Acridine Orange solution	10 s-2 min
(*e*)	Transfer sections to phosphate buffer, pH 6.0	1 min
(*f*)	Differentiate sections in calcium chloride solution	20 s
(*g*)	Transfer sections to phosphate buffer, pH 6.0	10 s
(*h*)	Mount sections wet and examine under fluorescent microscope	

Results

RNA : red

DNA: light green

Remarks

The authors state that formalin fixation should not be employed.

Method 15.7

DNA: Extraction

Reagents required

(*a*) Deoxyribonuclease

(*b*) 0.2 M Tris buffer, pH 7.6

Preparation of solution

Extraction of solution

0.2 M Tris buffer, pH 7.6	10 ml
Distilled water	50 ml
Deoxyribonuclease	10 mg

Method

(*a*) Bring test and control sections to water

(*b*) Place test section in extraction solution, control in tris buffer, pH 7.6 at 37°C, for 4 h

(*c*) Wash in running tap water

(*d*) Stain by method (Feulgen method) both sections

Results

Test section : DNA negative

Control section: DNA red

Method 15.8

RNA: Extraction

Reagents required

(*a*) Ribonuclease

(*b*) Distilled water

Preparation of solution

Ribonuclease	8 mg
Distilled water	10 ml

Method

(*a*) Bring test and control slides to water

(*b*) Place test slide in ribonuclease solution, control slide in distilled water at 37°C for 1 h (*c*) Wash in distilled water

(*d*) Apply Method 69, Methyl Green-Pyronin

Results

Test slide: RNA negative, DNA green

Controlled slide: RNA red, DNA green

Remarks

Nucleic acids may also be extracted by using perchloric acid. This is not as specific as enzyme digestion, but is a reliable technique and may be used on many occasions where extraction is required as considerable reduction is compared with enzyme digestion, the technique is given below.

Method 15.9

Extraction of nucleic acids with perchloric acid

Reagents required

Perchloric acid

Sodium carbonate

Preparation of solutions

(*a*)	Perchloric acid	2.5 ml
	Distilled water	47.5 ml
(*b*)	Perchloric acid	5 ml
	Distilled water	45 ml
(*c*)	Sodium carbonate	1 g
	Distilled water	100 ml

Method

To remove RNA only

(*a*) Bring sections down to water

(*b*) Place sections in 10 per cent perchloric acid (solution 2) at 4°C overnight

(*c*) Briefly rinse in distilled water

(*d*) Transfer to the sodium carbonate 5 min

(*e*) Wash in tap water

(*f*) Employ nucleic acid method

Remarks

To remove both RNA and DNA place sections in 5 per cent perchloric acid (solution 1) at 60°C for 30 min at stage (*b*) then continue method.

INDEX

A

B

C

D

E

F

G

H

I

K

L

M

T

U

V